Biodiversity of Parasite-fauna

Jammu, Kashmir and Ladakh Himalayas

Biodiversity of Parasite-fauna

Jammu, Kashmir and Ladakh Himalayas

R.C. BHAGAT
M.Sc. (Zoology), Ph. D. (Entomology)
Professor and Head
Post-Graduate Department of Zoology
Dean, Faculty of Science
The University of Kashmir, Hazratbal, Srinagar, Kashmir

Assisted by
POONAM BALA
B.Ed., M.Sc., M.Phil., Ph. D.
Lecturer
The Mallinsons Girls Higher Secondary School
Srinagar

PUBLISHING FOR ONE WORLD

NEW AGE INTERNATIONAL (P) LIMITED, PUBLISHERS
New Delhi • Bangalore • Chennai • Cochin • Guwahati • Hyderabad
Jalandhar • Kolkata • Lucknow • Mumbai • Ranchi
Visit us at www.newagepublishers.com

Published by New Age International (P) Ltd., Publishers
First Edition: 2008

ISBN (10): 81-224-2267-5

ISBN (13): 978-81-224-2267-2

Rs. 395.00

C-07-12-2163

Printed in India at Ramprintograph, Delhi.
Typeset through Author

PUBLISHING FOR ONE WORLD

NEW AGE INTERNATIONAL (P) LIMITED, PUBLISHERS
4835/24, Ansari Road, Daryaganj, New Delhi - 110002
Visit us at **www.newagepublishers.com**

PREFACE

This book aims to provide comprehensive account regarding biodiversity of many groups of animal parasites, belonging to Jammu, Kashmir and Ladakh Himalayan regions of Indian sub-continent. This is a first attempt to encapsulate vast published information on different groups of invertebrate parasites in this volume, such as protozoans, helminths, including plant-parasitic nematodes, annelids (leeches) and arthropods, published so far, lying scattered in numerous Indian and foreign publications like research journals, monographs, treatises, books and articles, etc.

In this work, the diversity of different kinds of parasites, with reference to their systematic survey and position, distribution, species richness and economic importance, parasitizing diverse range of hosts, have been dealt with lucidly and concisely. The different host groups, parasitized by parasite species, incorporated in the text are: Invertebrates (arthropods); Vertebrates– fishes, amphibians, reptiles, birds (wild and domesticated / poultry), mammals (wild and domesticated), man, plants (wild, domesticated crops and other economically important ones).

The text of this book mainly consists of 8 main Chapters besides, author and subject index, and bibliographical section. The indices have been designed to help readers to track specific subject information on parasite, including particular genus and researcher (author). The bibliographical section present at the end of the book is up-to-date, with citations and also encompassing titles of Ph.D. theses of J&K Himalayan regions, on general and economic parasitology, parasitic and entomophagous insects (parasites and parasitoids), and plant-parasitic nematodes.

Various Chapters of this book are laced with number of up-to-date checklists and catalogues, pertaining to different kinds of parasites, serving as quick reference for data on parasite/host species, systematic position of species nomenclatural changes, distribution and on authors (references).

Chapter 1, focusses on parasitic protozoans, infecting insects, fishes, birds, domestic mammals and man. The Platyheminthes (trematodes and cestodes), parasitic on molluscs, fishes, amphibians, birds, mammals (wild and domestic) and man, are discussed in Chapter 2.

Acanthocephalan parasites of vertebrates, with reference to fish hosts, are included in Chapter 3. An overview of parasitic nematodes (animals and plant) on different host groups, belonging to invertebrates (insects) as well as vertebrates in animals kingdom, and also parasitic on angiosperms in plant kingdom, is presented in Chapter 4.

Parasitic crustaceans (copepods), as ectoparasites of fishes and Acari (ticks and mites), parasites of wide spectrum of insects and vertebrate hosts, have been addressed in Chapter 5. The details on parasitic insects, *viz.* fleas, lice (chewing and biting), bugs, mosquitoes and dipteran flies, with their role as causing and vector of diseases in animals and human population, are provided in Chapter 6 & 7.

Chapter 8 is solely devoted to entomophagous insects, highlighting their faunistic/systematic survey, diversity, host range, species richness and their role as biological control agents of insect pests of agricultural, horticultural, silvicultural and forestry importance.

The present work has been conceived after keeping in view the impressive strides made during recent past on diverse animals parasite groups and the fact that the study of parasites is significant in the field of agriculture, public health, veterinary, pest control (biological) and, in biological diversity/biowealth knowledge of our region.

This volume, undoubtedly, cater to the growing needs of scientists, professionals and experts, conducting research in the field of Agriculture, Veterinary, Medical and Economic Parasitology, and Entomology besides, in the areas of other parasite groups, parasitizing not only invertebrate but also vertebrate hosts. Apart from this, it will be of immense help to students, scholars and researchers, in choosing the line/area of topic research, pertaining to different branches of general, economic/applied and arthropod parasitology, which awaits scientific investigation, either from inception or in-depth.

Lastly, I am thankful to M/S New Age International (P) Ltd., Publishers and the editorial Department for their keen interest in bringing out the book in a short span of period with a elegant form.

Suggestions and constructive criticisms for the improvement of the book are welcome.

R.C. BHAGAT

ACKNOWLEDGMENTS

At the outset, the author (R.C. Bhagat) expresses his sincere thanks to Hon'ble Vice-Chancellor, Kashmir University, Prof. Abdul Wahid Qureshi for providing unstinted moral support and encouragement. The working facilities provided by ex-Heads of P.G. Department of Zoology, Kashmir University., *viz.* Prof. A.R. Khan and Prof. M.Z. Chishti, are being greatly acknowledged.

Thanks are due to many colleagues of the author, Research Scholars, of Department of Zoology, Kashmir University, especially Dr. Fayaz Ahmad, Dr. Syeed Tanveer, Dr. Majidah Rasheed and, Dr. Mohd Niamat Ali., for providing relevant information, reprints of research papers, etc. used in the preparation of the present book.

Further, the author wishes to express his appreciation and thanks to his son, Rishabh Bhagat for his help in preparation of indices (author and subject), and sincere gratitudes due to his wife, Dr. (Mrs.) Poonam Bala, for her patience and great support in this project.

Special thanks are due to Mr. Javeid Ahmad Bhat, Universal Computers, Sadarbal, Srinagar, Mr. Firdous Ahmad, Sarai Bala, Srinagar, Mr. Irfan Ahmad Khan, Curator, Mr. Sajjad Naqash and Mr. Nazir Ahmad Rather, Department of Zoology, University of Kashmir, for their help in processing and computerizing the manuscript of this work.

The help rendered by Librarians and other staff of various National Libraries while consulting the literature, is gratefully acknowledged. These libraries are: Central Library, Indian Agricultural Research Institute, Library of Directorate of Plant Protection, Quarantine and Storage, Faridabad (Haryana). Library of Directorate of Health Services, New Delhi, National Institute of Science Communication and Information Resources, New Delhi, and Allama Iqbal Library, the University of Kashmir.

Lastly, the author dedicates the present book to his beloved parents.

CONTENTS

INTRODUCTION

Jammu and Kashmir State includes three Provinces, *viz.* Jammu, Kashmir and Ladakh, situated in North-west Himalayan region of Indian sub-continent. These areas are of paramount bio-geographical significance, not only with regard to their biological wealth but also for tremendous diversity of parasite-fauna. A great deal of knowledge pertaining to animal parasites, belonging to different groups of invertebrata, parasitizing diverse groups of hosts (invertebrates and vertebrates), is known, however, scattered and inaccessible, in different kinds of publications in India and abroad. This work is a pioneering attempt, to aptly present an up-to-date, lucid and concise account of biodiversity of parasitefauna of Jammu, Kashmir and Ladakh Himalayan regions.

The parasite is an organism that lives in or on the body of another organism or host. Parasites are metabolically dependent on their hosts and derive their nutrient requirements directly from them. The diverse parasite faunistic groups, known to exist in Jammu, Kashmir and Ladakh Himalayan regions are: parasitic protozoans; helminth parasites like trematodes, cestodes, acanthocephalan parasites, nematodes (animal and plant-parasitic); parasitic annelids (leeches); arthropods such as parasitic crustaceans (copepods), acari (ticks and parasitic mites), parasitic insects (fleas, lice, bugs mosquitoes and dipteran flies) and entomophagous parasitic insects.

The rich and diverse parasite-fauna of this region, has also shown great diversity with respect to their hosts. The various invertebrate group of hosts include annelids, arthropods, having umpteen insect species, belonging to different insect orders. The vertebrate host spectrum of parasite-fauna cover: fishes; amphibians (frogs and toads); reptiles (lizards and snakes); birds (wild, domesticated/poultry); mammals (wild and domesticated). Besides, human population, serving as host for number of ecto- and endoparasites.

The biodiversity of parasites, belonging to different faunistic groups, included in the text of present work, is briefly highlighted under the following heads:

(I) PARASITIC PROTOZOANS

Wide diversity of protozoan parasites, infecting invertebrates (insects), vertebrates (fishes, birds, domestic mammals) and human, has been reported from J & K, and Himalayan regions. Microsporidians (*Nosema* spp.), responsible for causing Nosema disease and Pebrine disease in honey bees (*Apis* spp.) and silkworm (*Bombyx mori*) respectively, are widely prevalent in this region. Fish protozoan diseases, *viz.* white-spot disease and Myxosporidian, are showing wide occurrence. Bird protozoans (haematozoa) show species richness (17 spp.), causing parasitaemia in 17 species/sub-species of birds, belonging to 11 families, occurring in wide localities and areas of Jammu & Kashmir.

Various protozoans, infecting domestic animals, *viz.* cow, bulls, equines, camel and yak, include *Babesia ovis*, *Theilaria annulata*, *Trypanosoma evansi*, *Eimeria* spp. and *Sarcocystis* sp. A total of 10 spp. of protozoan, is responsible for causing different diseases in human population. These include intestinal flagellates (4 spp.), amoebae (5 spp.) and Toxoplasmia (1 sp.), responsible for causing amoebiasis, enteritis, diarrhoea, giardiasis, toxoplasmosis, mildly and non-pathogenic diseases, in human population of this region.

(II) HELMINTH PARASITES AND PARASITIC ANNELIDS (LEECHES)

(a) Platyhelminthes (Trematodes and Cestodes)

Till to-date, a total of 105 spp. of trematode parasites, spread over 18 families, under Monogenea and Digenea, is known to parasitize diverse host animals like molluscs, fishes, amphibians, birds and mammals. During past few decades, a great deal of research work, has been done on taxonomy, pathogenicity and ecology of these parasites. The species richness of trematode parasites in vertebrate host groups, has been found highest in birds (49 spp.), followed by mammals (bat and domestic) (23 spp.), amphibians (frog and toad) (22 spp.) and fishes, having 9 spp.

About 48 spp. of tapeworms (cestodes), belonging to 25 genera and 11 families exist in this region. During recent years, detailed research studies have been done on faunistic diversity, prevalence/intensity of infection and population dynamics of about 11 spp. of cestodes, in different fish hosts, occurring in lakes (Dal, Anchar, Wular, Manasbal) river (Jehlum) and other water-bodies of Kashmir Valley. The vertebrate host-range of cestode parasites, with number of species, other than fish hosts are : amphibian (toad), 1 sp.; reptile (lizard), 1 sp.; birds (wild & domestic), 18 spp.; rat, 1 sp.; domestic animals (sheep, goat, bovids and yak), 11 spp. and human 2 spp.

In J & K, high prevalence of hydatid cyst disease of cattle and human due to cestode parasite, *Echinococcus* (*E. granulosus* and *E. multilocularis*), has been reported. In the recent years, detailed studies, including epidemiological on these parasites, have been accomplished. Hydatidosis (unicellular hydatid cyst disease) has shown wide diversity in infecting different sites/organs of human and cattles. Alveococcosis (multilocular or alveolar hydatid cyst disease), has also been found to exist in human population of Kashmir region.

(b) Acanthocephalan parasites

The diversity of 24 spp. of Acanthocephalan parasites, as endoparasites, damaging intestinal wall tissues of various vertebrate hosts by embedding the hooks present on the proboscis, has so far been reported Jammu and Kashmir Valley regions. In the recent years, thorough research studies have been accomplished on acanthocephalan parasites of fishes of Wular Lake. These studies cover their host spectrum and seasonal/population dynamics. Acanthocephalans have indicated wide incidence in fishes of Wular Lake. Acanthocephalan, *Pomphorhynchus kashmirensis* has shown maximum number of fish host species *i.e.* 9, followed by *Neoechinorhynchus manasbalensis* as 6 and *N. rutile*, having 2 fish host species.

(c) Animal nematodes

About 83 spp. of animal nematodes, distributed over 48 genera are known to infect invertebrates (insects) and vertebrates (fishes, amphibians, reptiles, birds, mammals and man), causing infection to different sites/organs and systems of these hosts. About nine species of nematodes have been found to infect domestic fowl (indigenous) of this region. Detailed studies on the prevalence and seasonal variation of nematode parasites of fowl of Kashmir, have also been done.

In-depth investigations on ovine nematodes, have indicated as many as 22 species of nematodes, infecting gastro-intestinal tract, abomasum and lungs of sheep, goat and lamb. Epidemiological studies done on ascariasis (animal) and on lungworms of sheep, have indicated wide prevalence of these diseases in Kashmir and Ladakh region. The important lungworms, infecting domestic ovines include *Dictyocaulus filaria, Mullerius* sp., *Prostostrongylus rufuscens* and *Varestrongylus pneumonicus*.

The incidence of nematodes, *Parafilaria* spp., causing haemorrhagic filariasis, has been reported in bovids from Jammu. Thelaziasis disease due to nematode parasite, *viz. Thelazia,* severely affecting eyes of cattle, has also been reported from Jammu. Indian Yak in Ladakh and Kashmir Stag (Hangul) in Dachigam (Kashmir), have been found to be infected by as many as 10 spp. and 7 spp. of gastro-intestinal nematodes respectively.

Several detailed research studies have been done on nematode parasites of human population of this region. The important parasites include *Ascaris lumbricoides*, Hookworm, *Enteriobus vermicularis* and *Trichuris trichuri*. High prevalence of human ascariasis in Kashmir has been reported, showing 80% in age group 6-10 years and less prevalence in high age group. Intestinal ascariasis is endemic in all parts of Kashmir. In addition to this, surgical heptobillary, gall bladder and pancreatic ascariasis, have also been widely reported from different region of Kashmir.

(d) Plant-parasitic nematodes

The data compiled during the course of present work, has shown that a total of 93 spp. of plant-parasitic nematodes are known to occur in Jammu, Kashmir and Ladakh region, causing damage to root, aerial and foliage parts of number of plant species, including valuable crops (fruits, cereals, vegetables) and other economically important plant species. Apart from this, 3 spp. of pathogenic nematodes of mushroom (*Agricus* spp.), have also been recorded from Kashmir.

(e) Parasitic leeches

About seven species of leeches as ectoparasites of warm and cold-blooded animals, feeding on blood, have been reported from this region. *Glossiphonia* has been found as a ectoparasite of snail (*Lymnea*), *Theromyzon,* parasitized waterfowl/bird and *Hirudinaria granulosa* is an important medicinal leech of this region.

(III) ARTHROPOD PARASITES

Arthropod parasites include parasitic copepods, Acari (ticks and parasitc mites) and entomophagous parasitic insects, exhibiting great diversity, species richness, wide occurrences and host range, in different regions and localities of Jammu, Kashmir and Ladakh Himalayan regions.

In general, parasitic insects (fleas, lice, bugs, mosquitoes and flies) are distinguished from entomophagous parasitic insects (parasites/parasitiods) in that their hosts are usually vertebrates, parasitic as adult stages, feed upon all stages of their host and do not kill their host. On the other hand, the entomophagous parasitic insects feed upon other insects and are parasitic only in immature stages, adults as free-living, known as parasitoids. The parasitoid larvae kill their hosts.

(a) Parasitic Crustaceans (copepods) and Acari (ticks and mites)

In this region, parasitic crustaceans are represented by three species of copepods, belonging to order Lernaepodoida. These are ectoparasites of fishes, remain attached to the gills and fins, and draw the nourishment from host's tissues.

A total of 47 spp. of ticks (soft and hard), has been found on diverse hosts: lizards, birds, bats, small burrowing mammals, dog, domestic cattle, beasts of burden and man. Ticks cause different diseases in these animals and man besides, play an important role as vector of human and animal diseases. Various diseases transmitted by different tick species of medical and veterinary importance include Tick-borne typhus, relapsing fever, Crimean-Congo haemorrhagic fever, avian spirochaetosis and anaplasmosis, bovine anaplasmosis, babesiosis and theileriosis.

About 21 spp. of mites, have so far been reported from various regions of J & K, parasitizing bat, burrowing mammals, birds, sheep and man, whereas 20 spp. of mites are parasitic on insects (houseflies, 2 spp. and honeybees, 18 spp.). 3 spp. of poultry mites show wide prevalence besides, psoroptic mange-causing mite, severely affecting sheep in the valley. Mite responsible for transmitting scrub typhus or tsutsugamushi fever in the human population, has been reported from Ladakh and Kashmir region. Species of mites, belonging to genera, *viz. Varroa* and *Tropilaelaps*, are most destructive parasites of honeybees (*Apis* spp.) in J & K regions.

(b) Parasitic Insects

(i) Fleas

29 spp. of fleas, have been found as haematophagous ectoparasites of birds (wild and poultry), bat, small burrowing mammals, dog, sheep, cattle and man. *Xenopsylla cheopis*, flea of medical importance, has been reported to transmit murine typhus disease. In addition to this, the wide prevalence of fleas of veterinary importance like stick-flea of poultry and dog flea, has been recorded from various localities and regions.

(ii) Lice (chewing and sucking)

25 spp. of sucking-lice, parasitizing small mammals, domestic mammals, dog, monkey and man, and 5 spp. of chewing lice, parasitizing poultry fowl, have so far been reported from this region. The chewing-lice of poultry importance include fluff-louse, *Gonicoeotes* (*G. gallinae, G. gigas*) and, *Gonicoeotes* (*G. gallinae, G. gigas*) and *Lepeurus capunis,* attacking chickens. Sucking-louse, *Pediculus humanus* is known to cause and transmit diseases in human, including earlier outbreaks of typhus disease in this region.

(iii) Parasitic bugs

The parasitic bugs belong to genus *Cimex* (bed bug). *C. hemipterus* and *C. lectularius*, suck blood of animal poultry and man. In addition to these, a rare bug, *Psammotettix striatus* was found to parasitize the high population of bat in Poonch (Jammu).

(iv) Mosquitoes

As many as 32 spp. of mosquitoes are known to occur in various localities, Districts and Provinces of J&K State. These include *Anopheles* (9 spp.), *Aedes* (7 spp.), *Finalaya* (1 sp.), *Culex* (14 spp.) and *Culiseta* (1 sp.). The main Anopheles mosquitoes known to transmit malaria disease in Jammu & Kashmir region are, *viz. Anopheles* (*Cilia*) *culicifacies* and A. (*C.*) *fluviatilis.*

(v) Dipteran flies (haematophagous and myiasis-causing)

37 species of haematophagous flies, feeding on blood of human and animals, have been found to occur in this region. These include 14 spp. of sandflies, 8 spp. of blackflies, 1 sp. of tabanid, 5 spp. of muscid and 9 spp. of pupipara. A few cases of leishmaniasis disease, transmitted by Sandfly *Phlebotomus major,* have earlier been detected in Jammu and Kashmir region. The pupipara of veterinary importance include sheep ked, *Hippobosca* spp. and *Lepotena capreoli*, parasitizing sheep, goat, cattle and dog.

In this region, blowflies or blue-green bottleflies (4 spp.), fleshflies (4 spp.), warbleflies (2 spp.), sheep or nostril maggotfly (1 sp.) and syrphidfly (1 sp.), have been found to cause myiasis. In myiasis, the larvae of the flies feed on host's tissues (healthy or necrotic), fluids (blood or other) or ingested food. The larvae of the blowflies and fleshflies now-a-days find use in the medical treatment of wounds, not responding to medication and this treatment is known as maggot therapy.

(c) Entomophagous insects (parasites and parasitoids)

The entomophagous parasites and parasitoids of this region show tremendous diversity, species richness and host range, belonging to insect order Diptera and Hymenoptera, Various dipteran entomophagous insects, existing in Kashmir are: canopids, muscids, sarcophagids and tachnids, covering a total of 20 spp., known to suppress the population of insect hosts like gypsymoth, bees, wasps and other unknown insect of various orders and families, found on dozens of host-plant species.

The hymenopteran parasites/parasitoids belonging to different families, with their number of species are: aphelinids (19 spp.); aphidiids (74 spp.); braconids (14 spp.); chalcids (6 spp.); cynipids (2 spp.); encyrtids (3 spp.); ichneumonids (38 spp.); megaspilid (1 sp.); mymarids (5 spp.); pteromalids (11 spp.); torymid (1 sp.) and trichogrammatids (2 spp.).

A wide family range, under different orders, including great number of host-insect and pest species, have been found to be attacked by aforementioned parasitoids, belonging to different families and insect orders. The various main host insect groups include aphids, beetles, moths, butterflies, sawflies, bugs, etc., found on more than hundred species of host plants, belonging to more than three dozen families. Many of entomophagous parasites and parasitoids play important role in suppression of insect pests of agricultural, silvicultural and forestry importance in many regions and localities, as efficient biological control agents. Some of the effective parasites/parasitoids might prove useful for exporting to other parts of India or world for, applied biological control programmes.

Chapter 1

PARASITIC PROTOZOANS

1.1 PROTOZOAN OF INVERTEBRATES (INSECTS)

Kaul (1987) investigated protozoan parasites of some invertebrates of Kashmir, however, the finding under M. Phil programme has not been published, some new records of parasites, with a few new species from hosts, *viz.* cockroach, molluscs, earthworm and grasshopper, have been included in the dissertation.

Protozoan, *viz. Nosema* spp., belonging to sub-phyulum Microspora (Microsporidian), Class Mircrosporea and Order Microsporida, show wide occurrence in J & K State, attacking Honeybees (*Apis* spp.) and Silkworm (*Bombyx mori*). The wide prevalence of *Nosema apis* and *N. bombycis* causing Nosema disease in honeybees and Pebrine in silkworm respectively. These protozoan species have been reported from Kashmir Valley by Shah & Shah (1988) and Sahaf (2002).

N. bombycis is very economically important parasite, it occurs in all tissues of the embryo, larvae, pupae and adult of *B. mori.* The infested silkworm shows numerous minute brownish-black spots, scattered over their bodies. The infected larvae is unable to spin cocoon and dies without pupation. Transovarian infection also occurs in this beneficial insect. *N. apis* causes great loss to workers frequently and, drones and queens also become susceptible to nosema disease. *N. apis* is found almost exclusively in the epithelial cells lining of host's gut and infection, unlike *N.bombycis* does not occur transovarially.

1.2 PROTOZOANS OF VERTEBRATES

The protozoans of vertebrates, include protozoans parasitic on fishes, birds, domestic mammals and human. These are discussed under the following heads:

1.2.1 Fish protozoans

In J & K, parasitic protozoan, *viz. Ichthyophthirius multifilis* Fouquet, belonging to Sub-phylum Ciliophora, Class Ciliata, Sub-class Holotrichidia, has been noticed to infect gills and skin of fish, *Aspidoparia morar* and *Triplophysa kashmirensis,* from Sehi Nalla (Jammu) and Telbal Nalla/Dal (Kashmir) respectively. This parasite was responsible for causing white-spot disease of these Fishes (Narian & Raina, 1992).

The heavy infection of ciliates on the gills and skin of the fishes has been observed and further the histo-pathological studies indicated gills in various stages of degeneration, with hyperplasia of connective tissues and deformity of respiratory folds.

A new myxoporidian, *viz. Myxidium cholai*, belonging to Subphylum Myxopora, Order Myxosporida, parasitic on fish, *Puntius chola* from Jammu (Ghoomansara Nallah), has been reported by Raina & Narain (1992). The infected gall-bladder of the fish was observed to be slightly elongated, with brownish bile.

1.2.2 Bird protozoans (haematozoa)

Detailed studies on blood protozoan parasites (Haematozoa) of some wild birds of Jammu and Kashmir, have been done by Nandi & Mandal (1978) and Nandi (1986). These authors made thorough blood examination of 39 species of birds and found parasitaemia in 17 species /sub-species, belonging to 11 families, *viz.* Accipitridae Capitoridae, Coryidae, Dicruridae, Fringillidae, Muscicapidae, Picidae, Pycnontidae, Strigidae, Sturnidae and Upupidae. The bird species, under these families showed infection due to 17 species under 4 genera of protozoan parasites, belonging to sub-order Haemosporina (sporozoa) and order Kinetoplastida (haemoflegellate). The checklist of blood protozoans and their host bird species is given below in Table: 1.1.

The species of Haemosporina include *Haemoproteus* (8 spp.) (Haemoproteidae), *Leucocytzoon* (7 spp.) (Leucocytozoidae) and *Plasmodium* (1 sp.) (Plasmodiidae). The Kinetoplastida is represented by *Trypanosoma avium* (Trypanosomidae). Insects, *viz.* mosquitoes black flies, hippoboscids, pupipara and other dipterous flies, act as intermediate hosts of these protozoans. The prevalence of parasitism due to protozoan parasites is believed to be as a result of insect vector activities.

High-incidence of protozoan infection in birds of Jammu and Kashmir, has been reported by the above cited authors. The various areas covered during the course of survey, included Anantnag. Ducksum, Doda, Jajjar Kotli, Khrew, Patnitop, Ramban and Udhampur. The haematozoa-fauna of J & K State showed species richness as well as diversity. It was revealed by Nandi & Mandal (1978) that higher incidence of haematozoa (26.8%) existed in J.&.K in comparison to that of Nasik and Ahmadnagar (Maharashtra) (13.3%) and Orissa (12.7%).

TABLE 1.1. Blood protozoan parasites and their host-bird species, with locality

S.No. (a)	Blood parasite/Host bird species (b)	Locality(ies) (c)
(1)	*Haematoproteus borgesi* *Jynx torquilla himalayana* Vaurie	DA, AN
(2)	*Haemotoproteus cellii* *Glaucidium cuculoides* (Vigors) *Strix aluco biddulphi* Seully	RA, DO, KH, AN
(3)	*Haemoproteus dicruri* *Dicrurus adsimilis* (Bechstein)	PA, UD
(4)	*Haemoproteus oryzivorae* *Myiophoneus caerulaus (Scopoli)*	JK, RA, DO

Table contd....

S.No. (a)	Blood Parasite/Host Bird Species (b)	Locality(ies) (c)
(5)	*Haemoproteus pastoris* *Acridotheres tristis tristis* finnaeus	RA, DO
(6)	*Haemoproteus thereicerycis* *Megalaima zeylanica*	JK
(7)	*Haemoproteus* sp.-1 *Garrulus lanceolatus* Vigors	PA, UD
(8)	*Haemoproteus* sp.- 2 *Butea vulpinus vulpinus* (Gloger)	JK
(9)	*Leucocytozoon danilewskyi* *Glaucidium cuculoides* (Vigors)	RA, DO
(10)	*Leucocytozoon dubreuili* *Pycnotus leucogenys humii* Oates	RA, DO
(11)	*Leucocytozoon fringinarum* *Carpodacus erythrinus* (Pallas) *Carpodacus erythrinus roseatus* (Blyth)	DA, ANDA, AN
(12)	*Leucocytozoon majoris* *Pycnonotus leucogenys humii* Oates	RA, DO
(13)	*Leucocytozoon squamatus* *Picus squamatus squamatus* (Vigors)	KH
(14)	*Leucocytozoon* sp-1 *Mycerobas aftinis* (Blyth)	DA, AN
(15)	*Leucocytozoon* sp-2 *Picus s. squamatus*	PA, UD
(16)	*Plasmodium relictum* *Muscicapa thalassina thalassina*–Swainson	PA, UD
(17)	*Trypanosoma avium* *Glaucidium cuculoides* (Vigors)	PA, UD

Key to Abbreviations: AN– Anantnag; DA– Ducksum; DO– Doda; JK– Jaggar Kotli; KH–Khrew; PA– Patnitop; RA– Ramban; UD– Udhampur.

1.2.3 Domestic animals protozoans

Various protozoans infecting domestic animals, *viz.* cow, bull, equines, camel and yak in Jammu, Kashmir and Ladakh region, studied in detailed, are as under:

1.2.3.1 *Babesia bovis* (= *B. argentina*)

B. bovis, belonging to subphylum Apicomplexa, Class Piroplasmea and family Babebsiidae, causes a disease known as babesiosis, widely distributed in the world, including Kashmir valley. The infection is acquired by female tick, *Boophilus,* which transmits the parasite transovarially to larval stages. The larval ticks inject sporozoites into blood of cattle, which invade erythrocytes. The infected erythrocytes clump together, resulting in blocking of capillaries, leading to brain damage and anoxia of internal organs. In Kashmir valley, Das & Shaw (1968) and Shaw (1980) gave earliest reports of the incidence of babesiosis in imported Australian Jersy cows and bulls.

1.2.3.2. *Theileria annulata* (Piroplasmea; Theileridae)

T. annulata causes disease in domestic animals known as tropical theileriosis, characterized by hypertrophy of the lymphoid tissues, fever, weight loss and haemolytic anaemia. The prevalence of theileriosis in cross-bred Jersey cow in Kashmir, has been shown by Shaw (1970, 80).

1.2.3.3. Trypanosomiasis, (Surra)

'Surra' disease is caused by protozoan, *Trypanosoma evansi* (Order Kinetoplastida, family Trypanosomatidae). Shaw (1980) reported the existence of this disease for the first time in Jammu province and its absence in Kashmir valley. The disease was prevalent among equine and camel in the plains of Kathua and Jammu (District). The field study has shown that flies, *viz. Tabanids* and *Stomoxys,* played role as mechanical vectors of 'Surra' disease. According to Cheng (1973), this disease is transient in cattle, fatal in camels, and is also known to parasitize donkeys, mules and dogs, in different parts of the world.

1.2.3.4 *Eimeria* spp. and *Sarcocystis* sp.

Ranga-Rao *et al.* (1994) have recorded two species of protozoan belonging to genus *Eimeria, viz. Eimeria brasiliensis* and *E. zurnii,* belonging to Sub-phylum Apicomplexa, Class Telosporea, Sub-class Coccidia and Order Eimerida. *Eimeria* spp. have been found to infect epithelial cells of some parts of intestine of Indian Yak (*Bos grunniens)* in Ladakh region. In addition to these parasites, above cited authors encountered Toxoplasmia, *viz. Sarcocystis* (large and small-sized) (Apicomplexa: Telosporea), in the form of cyst stages, present in the abdominal muscles of Yak.

1.2.4. Human protozoans

In J&K State, a total of 10 Species of protozoan parasites has shown occurrence in human populations in many localities. Of these parasites, 9 Species are as parasites of intestine. These are: flagellates

TABLE 1.2. Parasitic protozoans of Human population

S.No.	Name of protozoan species	Site of infection	Disease
(1)	*Chilomastix mesnile*	Intestine	Mildly pathogenic
(2)	*Embadomonas intestinalis*	-do-	Non-pathogenic
(3)	*Endolimax nana*	Large intestine	-do-
(4)	*Entamoeba coli*	-do-	-do-
(5)	*Entamoeba hartmanni*	Intestine	-do-
(6)	Entamoeba histolytica	Large intestine	Amoebiasis
(7)	*Enteromonas hominis*	Intestine	Non-pathogenic
(8)	*Giardia lamblia*	-do-	Enteritis, diarrhoea
(9)	*Iodamoeba butschlii*	-do-	Non-pathogenic
(10)	*Toxoplasma gondii*	Leucocytes & cells of reticuloendothelial system	Toxoplasmosis (causes sub-clinical infection)

(Mastigophora) - *Chilomastix mesnili, Embadomonas intestinalis, Enteromonas hominis,* and *Giardia lamblia*; amboebae (Sarcodina)- *Endolimax nana, Entamoeba coli, E. hartmanni, E. histolytica* and *Iodamoeba butschlii.* In addition to these, 1 species of Apicomplexa, Toxoplasmia (sporozoa), *viz. Toxoplasma gondii,* an intra-cellular parasite, is known to exist in this region.

The checklist of species of parasitic protozoans, occurring in human population in J.&.K, with their site of infection and diseases caused is presented in given on page 4. (Table 1.2).

Of the four species of intestinal flagellates, *Giardia lamblia* is pathogenic, causing enteritis and diarrhea, and rest of the three species are non-pathogenic. Of the five intestinal amoebae, reported from the human population of this region, *Entamoeba histolytica* is responsible for causing amoebiasis, it may be non-dysenteric infection, amoebic dysentery or may lead to abcesses in liver or lungs, etc. *Toxoplasma gondii* invades blood stream, leucocytes, tissues and cells of reticulo-endothelial system of human body. The disease caused by *T. gondii* is known as toxoplasmosis, studied in detail by Thokar (1986) in Kashmir Valley.

A detailed survey for the intestinal protozoans, based on stool samples of rural community in Lamberi village of district Rajouri was under taken by team from National Institute of Communicable Diseases (NICD), during the month of March, 1974. They reported a total of nine intestinal protozoan, these included *E. histolytica, E. coli, E. hartmanni, E. nana, I. butschlii, G. lamblia, C. mesnile E. intestinalis* and *E. hominis.*

The findings of NICD (Arora *et al.* 1976), revealed that 72.1% of the population of Lambari village was found to be infected with one or more parasite species. *E. coli* showed the highest percentage (48.6%) of existence in the community. The lowest percentage of intestinal protozoan encountered was 16.5 and 8.3 per cent in case of *G. lamiblia* and *E. histolytica* respectively.

Giardiasis in Kashmir Valley

Ahmad (1978), for the first time studied the incidence of *Giardia lamblia* in rural population of Kashmir. This author found that among other parasites *(Ascaris lumbricoides* and *Taenia saginata)* in rural population of Kashmir, *G. liamblia* accounted for 9.4% of total cases. Its prevalence was shown lower in males (8.1%) and higher in females (10.8%). The highest parasitic infection was observed in school-going age groups (5-15 years). The faeco-oral contact of poor hygiene and extensive contamination of soil, and water were believed to be the factors responsible for highest incidence of parasitic infection.

The incidence of giardiasis in urban areas (Srinagar) and surrounding areas, has been studied by Mir *et al.* (1979). These authors made detailed investigation on the incidence of *G. lamblia* parasite in Kashmiri children. For this purpose, 500 children from Srinagar and surrounding areas were selected for a period of one year. These childrens belonged to age group of 1-10 years. This study revealed total incidence of giardiasis 9.2% (46 cases) and giardiasis with ascariasis was 2.60% (13 cases). The maximum incidence was reported in children of age group 3-4 years. The personal hygiene of children and in sanitary condition were responsible for higher incidence of giadiasis in Srinagar. Recently, Khan *et al.* (2002) reported the prevalence of giardiasis in Kindergarten school children of urban Kashmir, covering 142 school of Srinagar city, indicated giardiasis in a total of 184 children (12.28%).

CHAPTER 2

PLATYHELMINTHES (Trematodes and Cestodes)

2.1. TREMATODE PARASITES

2.1.1. General account and species diversity

Trematode parasites, commonly referred to as flukes, belong to Phylum Platyhelminthes (flatworm) and Class Trematoda, characterized by the presence of organs of attachments (suckers). These are permanent parasites of internal organs of molluscs, harbouring larval asexual or occasionally sexual adult stage and of vertebrates, harbouring adult sexual stage, infecting different regions in the body cavity, organs and tissues.

In J & K State, trematode parasites are represented by two important Orders, *viz.,* Digenea and Monogenea. The monogenetic trematodes are usually ectoparasites of fish hosts, usually attached to the gill, scales and fins, etc., for example *Diplozoon* spp. Besides, one species, *Eupolystoma rajai*, has been reported to be true endoparasite in the urinary bladder/rectum of frog (*Rana* sp.). The monogeneids remain attached to the host surface by a large attachment organ, known as opisthaptor, species specific, provided with hooks and is present at posterior end of trematode. Usually, a pair of adhesive structures are located at the anterior end.

The digenetic trematodes of this region are very common and widespread parasites of various classes of vertebrates (fishes amphibians birds and mammals), occurring in adult as well as juvenile form, in every organ of their hosts. The digenetic trematodes are characterized by the presence of well developed sucker around the mouth, known as oral sucker. Besides, second ventral sucker or acetabulum is present in these types of trematodes. The suckers are involved in the attachment to the host as well as in locomotion. The organs of attachment in digenetic trematodes are poorly developed than those in monogenetic trematodes. Digenean parasites' development include at least two different hosts and several generations.

The first classified checklist of trematode parasites and their invertebrate host (snail) and vertebrate hosts (fishes, amphibians and birds) of Kashmir valley, has been given by Gupta (1967). This author has listed a total of 31 Species of trematodes, based on earlier reports from this region by various authors like Faust (1927), Jordon (1930), Mehra (1937), Kaw (1950), Fotedar (1964a, 65) and Fotedar & Raina (1964, 65). The listed species of trematode parasites included 9 spp., parasitizing 9 spp., of fishes, 17 spp. parasitic on 5 spp. of amphibians, 4 spp. parasitizing 4 spp. of birds. Besides these, 1 sp. of larval trematode infecting land snail, has been added in the checklist.

During the recent years, a great deal of research work has been done on taxonomy and ecology of trematode parasites of vertebrates like fishes, amphibians, birds and mammals, besides invertebrate (molluscan host), of J.&.K state. In this direction, an in-depth study of vast and scattered research publications, has revealed that till date, a total of 105 species of trematode parasites belonging to 56 genera, spread over more than 18 families, under Order Monogenea and Digenea, is known to occur in this region. The species diversity of trematode parasites of different host groups is given below in Table 2.1 and details regarding number of species of trematode parasites, belonging to different families and super-families, parasitizing different vertebrate groups is also given below in Table 2.2.

TABLE 2.1. Species diversity of trematode parasites of different vertebrate host groups of J & K

Host group	Genera	Number of speices	Families	Super family
Fishes	09	18	07	01
Amphibians (Frogs & Toads)	16	22	06	01
Birds (Wild & Poultry)	19	49	09	—
Mammals (Bat, domestic- & Yak)	12	23	05	—

Note: Two trematode, *viz. Neascus vetastai* from fish host and *Tetracotyle ranae* from frog, have only been assigned super family Diplostomomidea (after Niewiadomska, 2002), reported from this region as larval trematodes

TABLE 2.2. Number of species of trematode, under different families and super-families, parasitizing various vertebrate host groups in J & K

Trematode group	Vertebrate hosts Total number of species (Trematode)			
Order/Family (After Yamaguti, 1971)	FI	AM	BI	MA
(I) Monogenea				
(i) Diplozoidae	05	—	—	—
(ii) Polystomatidae	—	01	—	—
(II) Digenea				
(i) Allocreadiidae	05	—	—	—
(ii) Clinostomatidae	02	—	—	—
(iii) Cyclocoelidae	—	—	01	—
(iv) Dicrocoelidae	—	—	01	01
(v) Diplostomidae	01	01	03	—

Table contd...

Trematode Group	Vertebrate hosts Total Number of Species (Trematode)			
(vi) Echinostomatidae	—	—	29	—
(vii) Fasciolidae	—	—	—	03
(viii) Gorgoderidae	02	05	—	—
(ix) Lecithodendriidae	—	11	—	01
(x) Leucochloridiidae	—	—	01	—
(xi) Notocotylidae	—	—	09	—
(xii) Ophisthorchiidae	—	—	01	—
(xiii) Paramphi stomidae	—	01	17	
(xiv) Plagiorchiidae	01	02	02	01
(xv) Strigeidae	01	—	02	—
(xvi) Superfaily Diplostomoidae (After Nieniadomstka, 2002)	01	01	—	—
Total number of species	18	22	49	23

Abbreviation: FI = Fishes; AM = Amphibians (Frogs & Toad); BI = Birds (Poultry & Wild);MA = Mammals (Bat, domestic ruminants & Yak).

From Table 2.2, (p. 8), it is evident that as many as 49 species of trematode parasites, belonging to 9 families infect various wild and domestic birds, followed by 23 species, under 5 families, parasitizing mammalian hosts (bat, domestic ruminant and Yak, next 22 species, distributed over 7 families, infecting amphibian (toad and frogs), and lowest number as 18 species pertaining 8 families, attack fishes.

2.1.2 Molluscan and Fish trematodes

The larval stages of trematode genus *Brachylaima,* belonging to the family Brachylaimidae (Digenea), have been reported for the first time from Kashmir Valley, parasitic on various organs of land snail, *Buliminus dextrosinister* by Fotedar (1965). He also described the various larval stages of *Brachylaima,* such as sporocyst, cercaria and metacercaria.

The adults of Brachylaimid trematodes have been found in alimentary tracts of birds and mammals, rarely amphibians. The life-cycle of these trematodes include two intermediate hosts, both terrestrial molluscs. The larval stages: daughter-sporocyst are branched, cercaria with with rudimentary, stumpy tail, leaves first intermediate host; metacercaria encysted or not in kidney orpericardium of second intermediate host (Pojmanska, 2002).

Sharma *et al.* (1990) studied the post dormancy survival of molluscan host, *Lymnea auricularia* with trematode, *viz. Fasiola gigantica. F. gigantica* is a liver fluke of larger size and is found in sheep, goat and cattle in all parts of the world, including this region.

Till date, a total of 18 species of trematode parasites, belonging to 9 genera under 6 families of Sub-order Prosostomata (Order Digenea) and 1 family of Sub-order Polyopisthocotylea (Order Monogenea), besides larval from *Neascus vetastai,* under super-family Diplostomoidea of Digenca (after Niewiadomska, 2002), has been reported from Jammu and Kashmir region, infecting 18 species of fishes, distributed over 10 genera.

The Allocreadiids are represented by *Allocreadium fotedari, A. nemachilus, A. schizothoraxi, A. transversale* and *Crepidostomum indicum.* The species of Diplozoids (Monogenea) include *Diplozoon aegyptensis, D. cauveryi, D. guptai, D. kashmirensis* and *D. nipponicum.* The Plagiorchiid covers only one species, *viz. Astiotrema fotedari.* The Clinostomidids reported from this region are: *Clinostomum schizothoraxi* and *S. shafi.* The Gorgoderids are represented by two Species, *viz. Phyllodistomum loosi* and *P. megacotyle.* The family Diplostomidae and Strigeidae, is represented by *Neodiplostomum kashmirensis* and *Strigea annandalai* respectively.

In addition to the above trematode species, *Neascus vetastai,* reported by Kaw (1950) as larval form, has been included in the system of classification provided by Yamaguti (1971), in his monograph. However, in accordance with the recent classification, provided by Niewaidomska (2002), *N. vetastai* has been kept under super-family Diplostomoidea Poirier, 1886 of Sub-class Digenea.

The trematode parasites have been found to infect various organs and different parts of body of fish hosts in Jammu and Kashmir region. These are: Intestine, infected by *Allocreadium* spp., *Clinostomum shafi, Crepidostomum indicum, Diplozoom guptai, Phyllodistomum megacotyle* and *Strigea annadalai;* gills and gill cover by *Clinostomum schizotharaxi, C. shafi, Diplozoom* spp., *Neascus vetastai;* fins by *C. schizothoraxi* and *N. vetastai;* muscles by *C. schizothoraxi* and *Neodiplostomum kashmirensis;* eyes by *C. schizothoraxi;* viscera by *N. vetastai and* body cavity by *C. schizothoraxi.*

The trematode parasite spectrum of various host-fish species, is presented under the following lines:

Updated list of host-fish species of trematode parasites of J.&.K state:

1. ***Carassius carassius:*** *Clinostomum shafi, C. schizothoraxi, Diplozoon aegyptensis, D. cauveryi, D. guptai* and *D. kashmirensis*
2. ***Crossocheilus latius latius* (= *Crossochcilus letius*):** *Neodiplostomum kashmirensis*
3. ***Cyprinus carpio specularis:*** *C. schizothoraxi, D. Diplozoon kashmirensis, D, nipponicum*
4. ***Cyprinus* sp.:** *Diplozoon cauveryi*
5. ***Garra mullya:*** *Phyllodistomum megacotyla*
6. ***Glyptothorax* sp.:** *Clinostomum schizothoraxi*
7. ***Labeo dero:*** *Astiotrema fotedari*
8. ***Labeo* sp.:** *Clinostomum schizothoraxi, Doplozoon aegyplensis, D. cauveryi*
9. ***Nemacheilus rupecola:*** *Strigea anandalai*
10. ***Nemacheilus* sp.:** *Clinostomum schizothoraxi*
11. ***Oreinus* sp.:** *Diylozoon aegyptensis, D. cauveryi*
12. ***Schizothorax curvifrons (= Schizothorax hugelii) (= Schizothorax micropogon):*** *Clinostomum schizothoraxi, Diplozoon kashmirensis, Neodiplostomum kashmirensis, Neaseus vetastai*
13. ***Schizothorax esocinus:*** *Neascus vetastai, Clinostomum schizothoraxi, Diplozoon kashmirensis* and *Phyllodistonum loossi*

14. **Schizothorax niger:** *Clinostomuim shafi, Diplozoon kashmirensis C. schizothoraxi, D. aegyptensis, D. guptai, Allocreadium fotedari, A. transversale, Crepidostomum indicum*
15. **Schizothorax richardsonii (= Orienus sinuatus):** *Clinostomum schizothoraxi, Diplozoon kashmirensis, Neascus vetastai*
16. **Schizothorax sp.:** *Allocreadium schizothoraxi, Diplozoon kashmirensis and Phyllodistonum loosi*
17. **Triplophysa marmorata (= Nemacheilus kashmirensis):** *Allocreadium nemacheilus, Clinostomum schizothoraxi* and *Diplozoon kashmirensis*

Source: Ahmad & Chishti (1999, 2000); Chishti *et al.* (2002); Dhar (1978); Dhar & Kharoo (1985, 87); Dhar & Majidah (1987); Fotedar (1969); Fotedar & Parveen (1987); Faust ((1927); Gupta (1967); Kaw (1944, 50); Khan & Majidah (1994); Majidah & Khan (1998) and Pande (1938).

From the above given list, it is evident that maximum number of parasites i.e., 8 different species have been found to infect the fish, *viz. Schizothorax niger*. It is followed by 6 spp. of parasites in fish, *Carassius carassius,* 4 spp. each in *S. curvifrons* and *S. esocinus,* 3 spp. each in *S. richardsonii, Schizothorax* spp., *Triplophysa marmorata, Cyprinus, Carpio specularis, Laleo* spp., 2 spp. in *Oreinus* spp. Apart from these, 1 sp. of each parasite is known to infect fishes, *viz. Crossochelus latius, Cyprinus* spp., *Garra mullya, Glyptothorax, Labeo dero, Nemacheilus rupecola, Nemacheilus* sp. and *Triplophysa kashmirensis.*

Dhar & Kharoo (1986) made a detailed field study on trematode parasite, *viz. Clinostomum schizothoraxi* of fresh water fishes of Kashmir. These authors reported the existence of large number of metacercariae form of *C. schizothoraxi*, from gills, gill cover, fins and body surface of number of fresh water fishes, *viz. C. carpio specularis, S. niger, S. esocinus, S. curvifrons, S. richardsonii (=N. kashmirensis) and T. kashmirensis (=N. kashmirensis).*

Fish parasitization in detailed, covering helminthes, including Trematodes, *viz. C. schizothoraxi* and *D. kashmirensis,* from Wular Lake of Kashmir, was studied by Dhar & Majidah (1987). *C. schizothoraxi* was found from the skin and base of tail of *S. niger, S. esocinus* and *Nemacheilus* sp., whereas *D. kashmirensis* infected the gills of *S. niger, S. esocinus* and *C.c. specularis.*

Majidah & Khan (1998) studied the seasonal population dynamics of two trematode species of fishes, *viz. Diplozoon kashmirensis* and *Clinostomum schizothoraxi* in Wular Lake of Kashmir. These parasites were found to infect fishes like *S. niger, S. curvifrons, S. esocinus, Oreinus plagiostomus, Cyprinus carpio specularis* and *Nemacheilus kashmirensis.* The infra-population of parasite was shown to be relatively more during summer. The maximum infection was noticed in *S. esocinus, S. curvifrons* and *S. niger.*

In a comprehensive study of fish trematode parasite diversity in fresh water lakes of Kashmir, Chishti *et al.* (2002) found the occurrence of 5 spp. of trematodes, covering. 3 sp. of *Diplozoon* and 1 sp. each of *Clinostomum* and *Allocreadium* in fishes, *viz. Schizothorax, Oreinus, Labeo, Cyprinus* and *Carassius.* These authors found that digenetic trematode *(Clinostomum)* having a degree of host specificity and marked seasonal occurrences, whereas endoparasite trematode, *Allocreadium* showed occurrences in mid spring to late summer.

2.1.3 Amphibians (frogs and toads) and birds (wild and poultry) trematodes

In Jammu and Kashmir region, so far 22 spp. of trematode parasites, belonging to 16 genera, distributed over 6 families, under Order Monogenea and Digenea of class Trematoda, infecting frogs and toads, have been recorded. The Monogenea trematode included a single species, *viz. Eupolystoma rajai,* belonging to sub-family Eupolystomatinae and family Potystomatidae. The Digenea includes rest of the trematode species *i.e.* 21.

Toads *(Bufo* spp.) have been found to be parasitized by 11 different species of trematodes ,*B. viridis* by 9 spp., *B. latastei* and *Bufo* sp. by 1 sp. each, infecting intestine, rectum and urinary bladder of the host species. Of these Species of trematode, 10 spp. are known from Kashmir Himalayan region and 1 sp. from Jammu region. (Table 2.4, pp. 18-19).

Frog, *viz. Rana cyanophlyctis* has been reported to be infected by a total of 15 spp. of trematode parasites, with 18 spp. from Kashmir region and 7 spp. from Jammu. These trematode Species infect urinary bladder, instestine and lungs of *R. cyanophlyctis*. *Rana tigrina* from Jammu has been found to be parasitized by 3 spp. of trematodes, infecting intestine and urinary bladder of the host amphibian.

Apart from the above mentioned species of toads and frogs, *Rana* sp. from Poonch (Jammu) and *Bufo* sp. from Srinagar (Kashmir) have been found to be infected by trematodes, *E. rajai* and *Ganeo bufonis* respectively.

The most widely prevalent family of trematode parasite of amphibians in Jammu and Kashmir is Lecithodendriidae, incorporating 11 spp. of trematodes, distributed over 8 genera, followed Gorgoderidae, having 5 spp., Plagiorchiidae, having 2 spp. and rest of the families, *viz.* Polystomatidae (Monogenea), Paramphistomidae, Diplostomidae having one species each. Apart from these, 1 spp., *viz. Tetracotyle ranae,* has not been placed by Yamaguti (1958, 71) in any family, however, according to recent system of classification given by Niewiadomska (2002), genus *Tetraclyle* has been placed under Super-family Diplostomoidea (Table 2.4).

Different amphibian species of toad and frog, parasitized by different trematode parasite genera, with total number of species belonging to these are: *Bufo viridis,* infected with genus *Diplodiscus* (1 sp.), *Diplostomum* (1), *Gorgoderina* (1), *Loxogenes* (1), *Loxogenoides* (1); *Phyllodistomum* (1), *Plagiorchis* (1), *Pleurogenoides*(1), *Pleurorchis* (1); *Bufo latestei,* with *Plagiorchis* (1); *Rana cyanophlyctis,* with *Diplodiscus* (1), *Ganeo* (4 spp.), *Gorgoderina* (2 spp.) *Haematoloechus* (1), *Indopleurogenes* (1), *Loxogenes* (2), *Loxogenoides* (1), *Mehraorchis* (1), *Phyllodistomum* (1), *Prosotocus* (1), *Tetracotyle* (1); *Rana tigrina,* with *Diplodiscus* (1), *Ganeo* (1) and *Phyllodistomum* (1); *Rana* sp. with *Eupolystoma* (1); *Bufo* sp. with *Ganeo* (1), [See table 2.4].

Till-to-date, a total of 41 species of trematode have been reported to parasitize 18 different Species of birds (15 wild Species and 3 domesticated/ poultry), occurring in various regions and localities of Kashmir Valley. The majority of the Species of parasites, have been found in the birds of wetland habitats (Anchar, Hokersar, Dal Lake), and other localities of the valley like Litter, Dangerpora, Shalteng Qamarwari, Nowgam, Bijbehara, Zoorpora Bachpora, Ganderbal, and other regions and localities in district Srinagar. The 41 spp. of trematodes, are belonging to 19 different genera and 9 families, under sub-order Prosostomata and order Digenea, are listed in table 2.5, pp. 20-22.

The trematode parasites have been found to infect various regions in bird hosts' body such as: intestine (e.g. *Echinochasmus* spp, *Echinoparyphium* spp.); bile passage *(Lyperosomum kavini);* intestine and bile duct *(Pseudoechinostomum indicum);* cloaca *(Prosthogonimus cuneatus);* Caeca

(*Paramonostomum* spp.). Besides these sites of infection, trematode, *viz. Prosthogonimus putsehkowskii*, has been reported from the egg of domestic hen.

The various families of bird trematode parasites, incorporating different genera, with their total number of species are: Cyclococlidae, including genus *Typhloaoelum* (1 sp.); Dicrocoeliidae, genus *Lyperosomum* (1 sp.); Diplostomidae, 2 genera – *Posthodiplostomum* (1); *Prostnogonimum* (2), Echinostomatidae, 7 genera, *viz. Echinochasmus* (2), *Echinoparyphium* (8), *Echinostoma* (5), *Euparyphium* (1), *Hypoderacum* (1), *Paryphostomum* (4), *Pseudoechinostomum* (2), Leucochloridiidae; genus *Leucochloridium* (1), Ophisthorchiidae, genus *Opisthorahis* (1), Notoeotylidae, 3 genera – *Catatropis* (1), *Notocotylus* (3), *Paramonostomum* (3); Plagiorchiidae, genus *Plagiorchis* (2), Strigeidae, genera – *Apharyngostridea* (1), and *Cotylurus* (1),

The checklist of trematode parasites recorded from various bird species (wild, poultry/domesticated), is incorporated in the catalogue (Table 2.5, pp. 20-21) From this catalogue, it is evident that a total of 41 Species of trematodes, are known to parasitize about 18 Species of birds (wild and domestic). The wild birds are like Wigeon, Shoveller, Mallard (wild- duck), *Casarea*, Jungle-crow, Jackdaw, Coot, Moorhen, Crane, Bulbul, Kite, House sparrow, Carmorant, Little Greb (Dab-chick) and *Tringa*. Three Species of domesticated/poultry birds are: *Anas platyrhynchos domesticus* (domestic duck). *Anser anser domesticus* (Goose) and *Gallus gallus domesticus* (domestic bird). The parasite complex of these birds, is given in following catalogue:

Host-Bird Species—Trematode Species Catalogue

1. ***Anas penelope* (Wigeon)**: *Notocotylus attenuatus*
2. ***Anas platyrhynchos domesticus:*** *Cotylurus cornutus, Echinoparyphium agnatum, E. recurvatum, Echinostoma kashmirensis, E. revolutum, Hypoderatum conoideum* and *Pseudoechinostomum srinagarensis*
3. ***Anas platyrponchos* (Mallard /Wild Duck) (= *A. plathrynchos platyrhynchos*):** *Opisthorchis anasi, Pseudoechinostomum indicum*
4. ***Anser anser domesticus* (Goose)**: *Notocolylus babai, N. imbricatus*
5. ***Casarea ferruginea:*** *Echinostoma anseris casarcii*
6. ***Corvus macrorhynchos* (Jungle Crow)**: *Echinostoma revolutu*
7. ***Corvus monedula monedula* (Eastern Jackdaw) (= *Corvus monedula soemmeringi*):** *Lyperosomum kavini*
8. ***Fulica atra atra* (Coot/ kotru)**: *Paramonostomum fulicai*
9. ***Gallinula chloropus indica* (Indian Moorhen)**: *Echinoparyphium pindchi, Euparyphium pindchi* and *Leucochloridium gallinuli*
10. ***Gallus gallus domesticus* (Domestic fowl)**: *Catatropis kashmirensis, Echinochasmus coaxatus, Echinoparyphium kashmirensis, Echinostoma azerbaidjanicum, E. revolutum, E. turkestanicus, Paramonostomum elongatum, P. galli, Paryphostomum srivastavi, Prosthogonimus putschkowskii (egg parasite).*
11. ***Grus grus liffordi* (Crane)**: *Apharyngostridea garciai*
12. ***Malpastes haemorhous intermediates:*** *Plagiorchis bulbuli*
13. ***Milvus migrans lineatus* (Kite)**: *Posthodiplostomum milvi*
14. ***Passer domesticus* (House sparrow)**: *Plagiorchis tehri, Prosthogonumus cuneatus*
15. ***Phalocorocavax carbo sinensis* (Indian large Carmorant)**: *Notocotulus attenuatus, Paryphostonium mukhtarensis, P. radiatum* and *Paryphostomum* sp.

16. ***Podiceps ruficollis capensis* (Indian little Grel/Dabchick)**: *Echinochasmus fotedari and Euparyphium pindchi*
17. ***Tringa glaeola:*** *Echinostoma revolutum*

Source: Ahmad (2000); Ahmad & Chishti (1995, 97, 2000, 02); Chishti & Ahmad (1992); Chishti & Mir (1990); Chishti & Tanveer (1993); Dhar & Ahmad (1992); Dhar & Kharoo (1991); Fotedar (1964); Fotedar et al. (1970); Fotedar & Khateeb (1987); Fotedar & Raina (1964, 65); Khan (1987); Khan & Chishti (1982, 83, 84, 89); Kharoo & Dhar (1982); Khan & Ahmad (1992); Mehra (1937); Pandit *et al.* (1991); Raina (1971); Tanveer & Chishti (2001)

From the catalogue, given at page 13 and 14, it is evident that single trematode species parasitization, has been reported from the wild bird species, *viz. A. clypeata, A. penelope, C. ferruginea, C. macrorhynchos, C. m. monedula, F. a. atra, G. g. liffordi, M. h. intermedictus, M. m. lineatus, T. glaeola.* Two different species of trematode parasites, have been reported to infect birds like *A. platyrhynchos, Passer domesticus, Podiceps domesticus* and *P. r. capensis.* Three and four different species of trematode parasites in Kashmir valley, have been found in bird like *Gallinula indica,* and *Phalocorocarax carbo sinensis* respectively. Birds of poultry importance, *viz. A. b. domesticus* and *G. g. domesticus,* are affected by eleven and ten different Species of parasites respectively.

The trematodes of this region, having wide host-spectrum are: *Notocotylus attenuatus,* parasitic on birds like Wigeon and Indian large Carmorent. Euparyphium pindchi is also parasitic on two different bird hosts, *viz.* Indian Moorhen and Indian little Greb. *Echinostoma revolutum* has found different host birds– domestic duck, domestic fowl, jungle crow and *Tringa glaeola.*

Trematode Parasites of Domestic Fowl in Kashmir

In Kashmir Valley, the detailed studies on the prevalence, occurrence and seasonal variations of trematode parasites in common Kashmiri domestic fowl (*G. g. domesticus),* have been completed by: Khan & Chishti (1982), Fotedar & Khateeb (1987), Khan & Chishti (1989) and Pandit *et al.* (1991), indicating wide prevalence of trematode infections in domestic fowl.

A study conducted by Fotedar & Khateeb (1987) revealed the prevalence of infection by *Echinostoma azerbaidjanicus, E. revoltum, E. turkestanicus* and *Echinochasmus coaxatum* in the local fowl, collected from different places. These authors further reported that percentage of infection and parasite burden increased at optimum temperature, and during rainy season. The highest incidence of infection was recorded in the month of September, and lowest in the month of December and January.

Pandit *et al.* (1991) during the course of their study on the helminth parasites in indeginous fowl of Kashmir, have found the occurrence of only one Species of trematode *(Echinostoma revolutum),* accounted for about 1% of infection due to this parasites.

2.1.4. Mammals (bat,domestic animals and yak) trematodes

Two species of trematodes have so far been reported from Kashmir, infecting the intestine of bat, *Pipistrellus mimus* by Fotedar & Khan (1974).These parasites are: *Prosthodendrium (Paralecithodendrium) indicum* (= *Paralecithodendrian indicum*) Khan & Fotedar (1974) *(cf.* Yamaguti, 1971), belonging to sub-family Prosthodendriinae, family Lecithodendriidae (Prosostomata: Digenea) and *Plagiorchis (Plagiorchis) pipistrellicola* Fotedar & Khan, (1974a) belonging to the sub-family Plagiorchiinae, family Plagiorchiidae (Prosostomata: Digenea).

About 21 Species of rumen and liver trematode parasites of domestic ruminants (sheep, goat and cattle) and Indian Yak (domestic), are known to occur in various localities and regions of Kashmir and Ladakh. These species are belonging to 10 genera, spread over 3 families and 6 sub-families, under sub-order Prosostomata and order Digenea. The total number of trematode parasites species, belonging to various sub-families under different families are as: Paramphistomidae, sub-family Paramphistominae (8 spp.), Gastrothylacinae (3 spp.), Cladorchiinae (2 spp.), Orthocoeliinae (4 spp); Dicrocoeliidae, sub-family Dicrococliinae (1 sp.) and Fasciolide, sub-family Fascioliinae (3 spp.) (see table 2.6, pages 23-25).

The number of Species of trematode, infecting various domestic mammals like cattle (cow/ buffalo), sheep, yak and goat is 11, 10, 7 and 7 respectively. Trematode parasites of cattle in this region are, *viz. C. calicophorum, C. papillosum, Gastrothylax compressus, G. crumenifer, Olveria indica, Orthocoelium dicranocoelium, O. dinniki, O. gigantopharynx, O. indonesiens, Paramphistomum epiclitum,* and *P. gracilus* (see table 2.6, pages: 23-25). The trematode parasite species of sheep included *C. calicophorum, Ceylonocotyle scoloicoelium, Cotylophoron cotylophorum, Dicrocoelium* sp., *Fasciola* sp., *Gastrothglax compressus, G. crumenifer, Olveria thapari, Paramphistomum cervi and P. gracilis.*

Goats of this region, have been found to be infected by 7 Species of trematodes and these included. *C. caliciphorum, C. cotylophorum, dicrocoelium,* sp. *Fasiola* sp., *G. compressus, P. epiclitum* and *P. gracilis.*

Like common liver fluke, *Fasciola hepatica, Fasciola gigantica,* liver fluke of larger species is found in cattle, sheep and goat in all parts of India including Jammu & Kashmir State, causing much damage to these valuable domestic animals (See table).

The Indian Yak (*Bos grunniens*) in Ladakh region has been shown to be infected by 7 Species of trematode parasites by Ranga Rao *et al.* (1994). These included 3 species of flukes, infecting liver of Yak, *viz.* as *F. hepatica, F. gigantica* and *Paramphistamum explanatum (= Gigantocotyle explanatum)* and 4 Species of stomach fluke (amphistome), as *C. cotylophorum, G. crumenifer, Fischoederius cobboldi* and *P. cervi* (See table:2.6, page 23)

Prevalence of paramphistome trematodes in domestic ruminants of Kashmir

The earlier detailed study on the prevalence of paramphistomes in the domestic ruminants of Kashmir was done by Raina & Khan (1982). For this purpose, these authors conducted survey from October, 1980 to September, 1981. The collection of amphistomes of sheep and goat was made from slaughter houses present in Srinagar (Chota Bazar and Rainawari).In addition to this,the authors also made on the spot collection of amphistomes of the ruminants from far-flung slaughter houses, present in regions like Baramulla,Sopore,Anantnag and Handwara.The collection of amphistome parasites from cattle was made at Sheer, Heewan, Drangabal and Tral. The percentage of infection in sheep, goat and cattle was reported to be showing variation from month to month. The peak month of infection was found to be in May, in cattle and sheep, as 100% and 97.7% respectively. The peak infection in goat was in the month of June, showing 91.4% of infection.

Raina *et al.* (1987) completed thorough investigation on the paramphistomes, parasitizing the domestic ruminants of Kashmir. They reported and described eleven Species of paramphistomes from domestic ruminants (sheep, goat and bovines) collected from their rumina. The collection of trematode parasites was made from slaughter houses situated in various regions of Kashmir, *viz.* Srinagar, Pulwama, Anantnag, Verinag, Tral and Patten. The various genera and the number of Speices reported include *Calicophoron* (2 spp.), *Gastrothylax* (2 spp.) *Olveria* (1 sp.), *Orthocoelium* (4 spp.) and *Paramphistomum* (2 spp.). The termatode parasite, *Gastrothylax compressus* was reported to be as a commonly prevalent species of sheep and cattle in various areas of Kashmir, whereas *Oliveria indica* showed less common prevalence in this region.

TABLE 2.3. Trematode-fauna of fishes of Jammu and Kashmir

S. No.	Trematode species (a)	Location/s in fish (b)	Locality (ies) (c)	Author/s first reported parasite (d)
1.	*Allocreadium fotedari* Dhar & Kharoo	Intestine	Kashmir	Dhar & Kharoo (1985)
2.	*Allocreadium nemachilus*-Kaw	-do-	Bandipora, Ganderbal	Kaw (1950)
3.	*Allocreadium schizothoraxi*-Pande	-do-	Dal and Manasbal Lake	Pande (1938), Kaw (1944)
4.	*Allocreadium transversale*	-do-	-do-	Chishti et al. (2002)
5.	*Astiotrema fotedari* Dhar	-do-	Poonch (Jammu)	Dhar (1978)
6.	*Clinostomum schizothoraxi*-Kaw	Gills, muscles, caudal fin, eyes	Jehlum, Dal, Manasbal	Kaw (1950)
7.	*Clinostomum shafi* Ahmad-& Chishti	Intestine, gills	Anchar and Manasbal Lake	Ahmad & Chishti (2000)
8.	*Crepidostomam indicum*-Kaw	Intestine	Dal Lake	Kaw (1944)
9.	*Diplozoon aegytensis*-Fischthal *et* Kuntz	Gills	-do-	Ahmad & Chishti (1999)
10.	*Diplozoon cauveryi*	-do-	Dal and Manasbal Lake	Chishti et al. (2002)
11.	*Diplozoon guptai* Ahmad-& Chishti	Gills and Intestine	Anchar and Dal Lake	Ahmad & Chishti (1999)
12.	*Diplzoon kashmirensis*-Kaw	Gills	Dal, Manasbal, Anchar and Wular Lake	Kaw (1950)
13.	*Diplozoon nipponicum*-Goto	-do-	Kashmir	Fotedar & Parveen (1987)
14.	**Neascus vetastai* Kaw	Fin, gill cever, skin visceva	Jehlum river, Dal and Manasbal Lake	Kaw (1950)

Table contd....

S. No.	(a)	(b)	(c)	(d)
15.	*Neodiplostomum kashmirensis Faust*	Muscles	Jehlum river and Anchar Lake	Faust (1927) (cf. Gupta, 1967)
16.	*Phyllodistomum loosi* Kaw	Body cavity	Dal & Manasbal Lake and Jehlum (Shadipora)	Kaw (1950)
17.	*Phyllodistomum megacotyle*	Intestine	Poonch (Jammu)	Fotedar (1969)
18.	*Strigea annandalai* Faust	do	Kashmir	Faust (1927 (cf. Gupta, 1967)

The Systematic position of trematode species, cited above, under serial no. 1-13 & 15-18, based on Yamaguti (1958 & 71), is given below:-

Serial No./s (in the above table)	Order	Sub-order	Family	Sub-family
1–4	Digena	Prosostomata	Allocreadidate	Allocreadiinae
5	-do-	-do-	Plagiorchiidae	Astiotrematinae
6–7	-do-	-do-	Clinostomidae	Clinostominae
8	-do-	-do-	Allocreadiadae	Crepiadostominae
9–13	Monogenea	Polyopisthocotylea	Diplozoidae	
15	Digenea	Prosostomata	Diplostomidae	Diplostominae
16–17	-do-	-do-	Gorgoderiadae	Phyllodistominae
18	-do-	-do-	Strigeidae	Strigeinae

* Indicates that Yamaguti (1971) has not assigned systematic position to trematode (serial no. 14), *viz. Neascus vetastai* reported by Kaw (1950) from fishes of Kashmir as larval trematode and adult of this species is unknown in other places. However, according to new system of classification of trematoda (Class), Niewiadomska (2002) has placed genus *Neascus* under Super-family Diplostomoida of Sub-Class Digenea.

TABLE 2.4. ***Trematode-fauna, parasitizing amphibians (Frogs and toad) in Jammu and Kashmir***

S.No.	Trematode species	Amphibian species	Location of parasite	Locality
1.	*Diplodiscus amphichrus* (=*Diplodiscus mehrai*)	*Bufo viridis, Rana cyanophlyctis, R. tigrina*	Rectum, Intestine	Srinagar, Poonch and Jammu
2.	*Diplostomum bufonis*	*Bufo viridis*	Intestine	Srinagar (Kashmir)
3.	*Eupolystoma rajai*	*Rana* sp.	Urinary bladder, Rectum	Poonch (Jammu)
4.	*Ganeo bufonis*	*Bufo* sp.	Intestine	Srinagar
5.	*Ganeo srinagarensis*	*Rana cyanophlyctis*	-do-	-do-
6.	*Ganeo tigrinus*	*R. cyanophlyctis, R. tigrina*	-do-	Jammu
7.	*Gorgoderina elliptica*	*R. cyanophlyctis*	-do-	-do-
8.	*Gorgoderina vitelliloba* (= *Phyllodistomum frequentum*)	*R. cyanophlyctis, B. viridis*	Urinary bladder	Kashmir
9.	*Haematoloechus almorai* (= *Pneumonoeces almorai*)	*R. cyanophlyctis*	Lungs	-do-
10.	*Indopleurogenes yamaguti*	*R. cyanophlyctis*	Intestine	Jammu
11.	*Loxogenes jammuensis*	*Rana cyanophlyctis*	Intestine	Jammu
12.	*Loxogenes kashmirensis* (= *Nenimandijea kashmirensis*)	*R. cyanophlyctis, B. viridis*	-do-	Kashmir
13.	*Loxogenoides bicolor* (= *Loxogenes bicolorr*) (= *Loxogenes arcanum*)	*R. cyanophlyctis, B. viridis*	-do-	-do-
14.	*Mehraorchis ranarum*	*R. cyanophlyctis*	-do-	Jammu
15.	*Phyllodistomum almorai*	*do*	Urinary bladder	Kashmir
16.	*Phyllodistomum shandrai*	*Rana tigrina*	-do-	Jammu
17.	*Phyllodistomum* sp.	*Bufo viridis*	-do-	Srinagar
18.	*Plagiorchis himalayai* (= *Lepoderma himalayai*)	*B. viridis*	Intestine	Sriangar

Table contd....

19.	*Prosotocus himalayai* (=. *Prosotocus partapus)* (= *Prostocus kashabia)*	*Rana cyanophlyctis*	Intestine	Jammu & Kashmir
20.	*Pleurogenoides buforis*	*Bufo viridis*	-do-	Srinagar
21.	*Pleurorchis nensie*	*-do-*	-do-	Kashmir
22.	*Tetracotyle ranae**	*Rana cyanophlyctis*	-do-	-do-

Source: (Serial No. 1–22): Bhalerao (1937); Fotedar (1959, 67, 69 70 & 71); Gupta (1967); Jordan (1930); Kaw (1943, 45 & 50); Pande (1937); Singh *et.al.* (2001); Tickoo (1970).Systematic position of amphibian treematode species, cited in the above table (Serial no. 1–21), in accordance with the scheme of classification provided by Yamaguti (1958, 71), is presented as under:

Serial no./s above table)	Order	Sub-order	Family	Sub-family
3	Monogenea	–	Polystomatidae	Eupolystomatinae
1	Digenea	Prosostomata	Paramphistomidae	Diplodiscinae
2	-do-	-do-	Diplostomidae	Diplostominae
7–8	-do-	-do-	Gorgoderidae	Gorgoderinae
15–17	-do-	-do-	-do-	Phylklodistominae
18	-do-	-do-	Plagiorchiidae	Plagiorchiinae
9	Digenea	-do-	do	Haematoloechniae
4–6	-do-	-do-	Lecithodendriidae	Ganeoginae
10	-do-	-do-	-do-	Pleurogeninae
11–13, 20–21	-do-	-do-	-do-	Pleurogenetinae
14, 19	-do-	-do-	-do-	Prosotocinae

Serial No. 22 (*) has been included in the monographs of Yamaguti (1958, 1971), however, this species of trematode was not assigned any family/ subfamily in his system of classification, because of being a larval trematode. In the latest version of systematics of trematode, given by Niewiadomska (2002), has placed genus *Tetracotyle* in the Superfamily Diplostomoidea under Sub-Class Digenea of class Trematoda.

TABLE 2.5. Checklist of Trematode-fauna, parasitizing birds (wild and poultry) in Kashmir Valley.

S.No.	Trematode Species	Author(s) first recording
1.	*Aphayngostridea garciai*	Dhar & Kharoo (1991)
2.	*Catatropis kashmirensis* Khan & Chishti	Khan & Chishti (1989)
3.	*Cotylurus cornutus* Rudolphi	Dhar & Ahmad (1992)
4.	*Echinochasmus coaxatus* Dietz	Fotedar & Khateeb (1987)
5.	*Echinochasmus (E.) fotedari* Chishti & Mir.	Chishti & Mir (1990)
6.	*Echinoparyphium angatum* Dietz	Ahmad & Chishti (1997)
7.	*Echinoparyphium chinense* Lee *et al.*	Khan & Ahmad (1992)
8.	*Echinoparyphium cleric* Skrybin	Dhar & Ahmad (1992)
9.	*Echinoparyphium distinctum* Dietz	Ahmad & Chishti (1997)
10.	*Echinoparyphium haneyanum* Johnston	-do-
11.	*Echinoparyphium kashmirensis* Chishti & Tanveer	Chishti & Tanveer (1992)
12.	*Echinopanyphium pindchi* Dhar & Kharoo	Dhar & Kharoo (1991)
13.	*Echiniopatyphium recurvatum* (Linstow)	Ahmad & Chishti (1997)
14.	*Echinostoma anseris casarcii* Dhar & Kharoo	Dhar & Kharoo (1991)
15.	*Echinostoma azerbaidjanicum* Kasimov	Fotedar & Khateeb (1987)
16.	*Echinostoma kashmirensis* Chishti & Ahmad	Chishti & Ahmad (1992)
17.	*Echinostoma revolutum* (Froelich)	Khan & Chishti (1982a)
18.	*Echinostoma turkestanicus* Kurova	Fotedar & Khateeb (1987)
19.	*Euparyphium pindchi* Khan & Chishti	Khan & Chishti (1984)
20.	*Hypoderacum conoideum* (Block)	Khan & Ahmad (1992)
21.	*Leucochloridium gallinuli* Kharoo & Dhar	Kharoo & Dhar (1982)
22.	*Lyperosomum kavini* Fotedar & Raina	Fotedar & Raina (1965)
23.	*Notocotylus attenuatus* Rudolphi	Ahmad & Chishti (1995)
24.	*Notocotylus babai* Bhalerao	-do-
25.	*Notocolyus imbricatus* (Looss)	-do-
26.	*Opisthorchis anasi* Khan	Khan (1987)
27.	*Paramonostomum elongatum* Yamaguti	Tanveer & Chishti (2001)
28.	*Paramonostomum fulicai* Bangh	-do-
29.	*Paramonostomum galli*	-do-

Table contd....

S.No.	Trematode Species	Author(s) first recording
30.	*Paryphostomum mukhtarensis* Ahmad	Ahmad (2000)
31.	*Paryphostomum radiatum* Dietz	-do-
32.	*Paryphostomum srivastavi* Chishti & Tanveer	Chishti & Tanveer (1993)
33.	*Paryphostomum* sp.	Ahmad (2000)
34.	*Plagiorchis bulbuli* (Mehva) (= *Lepoderma bulbuli* Mehva)	Mehra (1937)(*cf.*Yamaguti 1958, 71)
35.	*Plagiorchis (P. tehri* Khan & Chishti)	Khan & Chishti (1983)
36.	*Posthodiplostomum milvi* Fotedar & Raina	Fotedar & Raina (1964)
37.	*Prosthogonimus (M.) cuneatus* (Rudolphi)	Fotedar *et al.* (1970)
38.	*Prosthogonumus putschkowskii* Skrjabin	Raina (1971)
39.	*Pseudoechinostomum indicum* Mehva	Chishti & Ahmad (1992)
40.	*Pseudechionostomum srinagarensis*– Dhar & Ahmad	Dhar & Ahmad (1992)
41.	*Typhlocoelum indicum* Fotedar	Fotedar (1964a)

The location of tematode parasite species (serial 1–41) in different host birds (wild & domestic), is given as under:

Serial No./s (Table 2.5)	Location of parasite in host body
1–21, 23–23, 28–36, 40–41	Intestine
22	Bile passage
27	Caecum
37	Cloaca
38	Egg
39	Intestine & bile duct.

TABLE 2.5a. Classification of bird trematode species, cited in the table 2.5, pages 20–21 (serial 1–41, based on the system of classfication provided by Yamaguti (1958–1971).

Serial No. (s) (in table 2.5)	Order	Sub-order	Family	Sub-family
1	Digenea	Prosostomata	Strigeidae	Strigeinae
3	-do-	-do-	-do-	-do-
2, 23–25, 27–29	-do-	-do	Notocotylidae	Notocotylinae
4–20, 30–33, 39–40	-do-	-do-	Echinostomatidae	Echinostomatinae
34–35	-do-	-do-	Plagiorchiidae	Plagiorchiinae
21	-do-	-do-	Leucochloridiidae	Leucochloridiinae
	-do-	-do-	Dicrocoeliidae	Dicrocoeliinae
22	-do-	-do-	-do-	Lyperosomini
26	-do-	-do-	Ophisthorchiidae	Ophisthorchiinae
	-do-	-do-	Diplostomidae	Diplostominae
36–38	-do-	-do-	-do-	Diplostomini
41	-do-	-do-	Cyclocoelidae	Typhocoelinae

TABLE: 2.6. Checklist of Rumen and Liver trematode parasites of domestic ruminants (sheep, goat and cattle) and Indian Yak of J & K State

S. No.	Trematode species	Host	Location	Locality	Author (s) reporting first
(a)	(b)	(c)	(d)	(e)	(f)
1.	*Calicophoron calicophorum* (Fischoeder) (= *Paramphistomum skrijaabini* Popova)	SH, GO, CA SH	R R	K K	Bali (1971) Bali & Fotedar (1972) (*cf.* Raina *et al.* (1987)
2.	*Calicophoron papillosum*	CO, BU	R	K	Raina *et al.* (1987)
	(Stiles et Goloberger)				
3.	*Ceylonocotyle scoliocoelium*	SH	R	JK	Bali & Fotedar (1972)
	(Fischoeder)				
4.	*Cotylophoron cotylophorum* (Fischoeder)	SH, GO, YA	R, S	K, L	Bali (1976), Ranga Rao *et al.* (1994)
5.	*Dicrocoelium* sp.	SH, GO	GI	K	Makhdoomi *et al.* (1995)
6.	*Fasciola hepatica*	YA	LI, GB	L	Ranga Rao *et al.* (1994)
7.	*Fasciola gigantica*	YA	LI, GB	L	-do-
8.	*Fasciola* sp.	SH, GO	GI	K	Makhdoomi *et al.* (1995)
9.	*Fischoederius cobboldi*	YA	S	L	Ranga Rao *et al.* (1994)
10.	*Gastrothylax compressus* Brandes	GO	R	K	Sey (1983), Raina et al. (1987)
	(=. *Gastrothiylax indicus* Dutt)	DRU, SH	R	K	-do-
	(=. *Gastrothylax orientalis* Bambroo)	DRU, SH	R	K	Bambroo (1970)
	(=. *Gastrothylax glandi formis-*	DRU, SH	R	K	Bambroo (1970)
	Bambroo)	DRU, YA	R,S	K,L	

Table contd....

(a)	(b)	(c)	(d)	(e)	(f)
11.	*Gastrothylax crumenifer* (Creplin)	DRU, YA	R, S	K, L	Bali (1976) Ranga Rao *et al.* (1994)
12.	*Olveria indica* Thapar & Sinha	BU	R	K	Raina *et al.* (1987)
13.	*Olveria thapari* Bali & Fotedar	SH	R	K	Bali & Fotedar (1972 a)
14.	*Orthocoelium dicranocoelium* (Fishoeder)	DRU	R	K	Raina *et al.* (1987)
15.	*Orthocoelium dinniki* Eduardo	DRU	R	K	-do-
16.	*Orthocoeium gigantophanynx* (Schad *et al.*)	DRU	R	K	-do-
17.	*Orthocoeoum indonesiense* Eduardo	DRU	R	K	-do-
18.	*Paramphistomum cervi* (Zeder)	SH, YA	R	L, L	Bali (1976), Ranga Rao *et al.* (1994)
19.	*Paramphistomum epiclitum*- Fischoeder	SH, GO, CA	R	K	Raina *et al.* (1987)
20.	*Paramphistonum explanatum* (= *Gigantocotyle explanatum*)	YA	S	L	Ranga Rao *et al.* (1994)
21.	*Paramphistomum gracilis* Fischoeder	DRU, SH, GO, CA	R	K	Raina *et al.* (1987)
22.	Excysted amphistome (gastrothylaiid metacercariae)	DRU	R	K	Raina & Khan (1983)

Abbreviations used: BU = Buffalo; CA = Cattle; CO = Cow; DRU = Domestic ruminants; GB = Gall bladder; GI = Gastro-intestine; GO = Goat; JK = Jammu & Kashmir; K = Kashmir; L = Ladakh; LI = Liver; R = Rumen; S = Stomach; SH = Sheep; YA= YA = Yak (Indian)

TABLE 2.6a. Systematic position of trematode species of domestic cattle, given in the table 2.6, serial no. 1-21 pages 23-24

Serial No. (s)	Order	Sub-order	Family	Sub-family	Tribe
1–2, 4	Digenea	Prosostomata	Paramphistomdae	Paramphistominae	Paramphistomini
3	-do-	-do-	-do-	-do-	Ceylonocotylini
10–11	-do-	-do-	-do-	Gastrothylacinae	-
12–13	-do-	-do-	-do-	Cladorchiinae	Olveriini
14–17	-do-	-do-	-do-	Orthocoeliinae	-
18–21	-do-	-do-	-do-	Paramphistominae	-
9	-do-	-do-	Gastrothylacidae	Gastrothylacinae	-
5	-do-	-do-	Dicrocoeliidae	Dicrocoeliinae	Dicrocoeliini
6–8	-do-	-do-	Fasciolidae	Fascioliinae	-

For classification of parasites (trematodes) in the tables 2.6 and 2.6a, Yamaguti (1971), has been consulted.

2.2 CESTODE PARASITES

2.2.1 General account

Cestodes or tapeworm parasites belong to Class Cestoda of Phylum Platyhelminthes and possess an elongated tape-like body, dorsoventrally flattened, usually white in colour, however, some species show yellow or gray colouration. Cestodes are endoparasites of various Vertebrates (fishes, amphibians, reptiles, birds and mammals, including man).

The body of the typical cestode parasite is divided into three regions: (i) Scolex, the organ of attachment; (ii) the neck region, situated immediately posterior to scolex and (iii) strobila, which constitutes the main portion of the body and is made up of a chain of proglottids. The most anteriorly present proglottids, are generally immature and proglottids posterior to immature ones are sexually mature.

During the life-cycle of cestodes, one or two, sometimes even more intermediate hosts are required. The life-cycle pattern and larval habitat show wide variation in different groups of cestodes. The larvae are known to parasitize any organ or part of vertebrate as well as invertebrate host.

In this region, a total of 48 species of cestodes, belonging to 25 genera, are known to be parasitic on various vertebrate groups. These species come under two sub-classes: Cestodaria (Order, Caryophyllidea); Eucestoda (Order, Pseudophyllidea, Proteocephalidea and Cyclophyllidea). The four orders of class Cestoda, incorporate a total of eleven families and nine distinct known sub-families (Table:).

In Cestodaria group, the cestodes are not divided into segment and contain only one set of reproductive organs; the scolex is lacking; presence of 10-hooked larva (decanth). In Eucestoda, the body is divided into segments or proglottids, each containing a set of male and female reproductive organs; scolex is usually present; presence of 6-hooked.

The Caryophyllid is represented by one family, *viz.* Caryophyllaeidae, including one genus, *viz. Adenoscolex*, parasite of fishes various localities and areas of Kashmir. The Caryophyllaeid cestodes show the presence of single set of reproductive organs and utilize aquatic oligochaetes as intermediate host. The adults are parasitic in fishes. These are usually small unsegmented tapeworms.

The Pseudophyllid cestodes are represented by two families, *viz.* Bothriocephalidae and Diphyllobothriidae, including genus *Bothriocephalus* and *Ligula* respectively. These cestodes are comparatively small, and chiefly parasites of fishes in Kashmir region. Pseudophyllid parasites are characterized by the presence of true sucking grooves (bothria) on the scolices.

The Proteocephalid cestodes include two genera – *Proteocephalus* and *Gangesia,* belonging to the family Proteocephalidae, parasitizing fishes of Jammu and Kashmir region. These cestodes are distinguished from one another by the arrangement of reproductive organs inside each proglottid. The eggs of Proteocephalids are embryonated when laid and hatch on ingestion by the suitable intermediate host, *viz.* copepod.

A total of twenty genera, have been covered under the order Cyclophyllidea, distributed over seven families and eight sub-families. These include most common tapeworms that are parasitic on birds, amphibians, reptile, mammals, including man, and are of great veterinary and medical importance.

2.2.2 Fish, amphibian, reptile and bird cestodes

2.2.2.1 Fish cestodes

The earliest systematic survey for the cestode parasites of fishes of J&K State, have been conducted by Datta (1936), who for the first time reported the existence of cestode, *viz. Ligula intestinalis* (Diphyllobothriidae) in the intestine of fish *(Schizothorax)* of Dale lake. Later, Fotedar (1958) gave a description of new genus/species of Caryophyllaeid, *Adenoscolex oreini,* infecting intestine of *Schizothorax richardsonii (=Oreinus sinuatus)* from Kashmir region.

In the later years, more cestode parasites such as: *Bothriocephalus acheilognathi* (Bothriocephalidae); *Gangesia fotedari, G. lucknowi, Gangesia* sp., *G. jammuensis, G. kashmirensis, Proteocephalus kashmirensis* and *P. torulosus* (Proteocephalidae: Proteocephalinae), were recorded from the intestine of number of fish host species,existing in different water bodies of various localities/areas of Jammu and Kashmir (Dhar & Fotedar, 1980; Dhar & Raina, 1983; Dhar & Majidah, 1983 and Ara & Ahmad, 2000).

In the recent years, a number of detailed studies have been accomplished on fish parasitization by different cestode parasites, occurring in different lakes of Kashmir, covering diversity, occurrence, seasonal variation, with influence of season on the intensity of infection due to parasite species, has been accomplished by Dhar & Majidah (1987), Dhar & Peerzada (1992), Khan & Majidah, 1993 (1994), Majidah & Khan (1998) and Chishti *et al.* (2002).

Till-to-date, the available published data on the occurrence of Cestode species infecting intestine of different host fishes of J&K State, have revealed that a total of about eleven species of cestode parasites, distributed over five different genera, belonging to four families: Caryophyllidae, Bothriocephalidae, Diphyllobothriidae and Proteocephalidae, are known to exist in various localities of Jammu and Kashmir region.

The cestode parasite-complex of various fish host species, occurring in lakes (Dal, Anchar, Wular, Manasbal), Jehlum river and other water-bodies, in the light of recent revised nomenclature changes, with respect of fish species, is listed as under:

S. No.	Fish Species	Cestode Parasite Species
1.	***Carassius carassius***	*Adenoscolex oreini*
2.	***Cyprinus carpio communis***	*Bothriocephalus acheilognathi*
3.	***Cyprinus carpio specularis***	*A. oreini, Gangesia* sp.
4.	***Glyptothorax***	*Gangesia fotedari, Gangesia* sp. *G. kashmirensis, G. lucknowi*
5.	***Schizothorax esocinus*** ***(=. Schizothorax hugelii)***	*A. oreini, Gangesia* sp.
6.	***Schizothorax richardsonii*** ***(=. Oreinus sinuatus)*** ***(=O. plagiostomus)***	*A. oreini, Gangesia* sp.
7.	***Schizothorax curvifrons***	*A. oreini, Gangesia* sp.
8.	***Schizothorax niger***	*A. oreini, Gangesia* sp.
9.	***Triplophysa kashmirensis*** ***(=. Nemacheilus kashmirensis)***	*A. oreini, Gangesia* sp. and *Protocephalus kashmirensis)*
10.	***Schizothorax* sp.**	*Adenoscolex* sp., *Bothriocephalus* sp. *Ligula intestinalis*
(11).	***Wallago attu***	*Gangesia jammuensis*

From above, it is clear that the fish parasitization, with decreasing order of number of parasite species in various fish host species is as: *Glyptothorax,* having infected by highest number of parasites as five, followed by *Schizothorax* sp. and *Triplophysa kashmirensis,* by three parasite species each *C. c. specularis, S. esocinus, S. richardsonii, S. curvifrons, S. niger,* infected with 2 species each and rest of the fishes like *Carassius carassius, C. c. communis* and *Wallago attu,* infected by only one species each of cestode parasite.

Detailed investigations on the seasonal population fluctuation of *Adenoscolex oreini,* infecting various fishes, *viz. S. niger, S. esocinus* (= *S. hugelii*), *S. curvifrons* and *Schizothorax richardsonii* (=*Oreinus sinuatus*), *O. plagiostomus* in Wular lake, have been carried out for the first time by Dhar & Peerzada (1992).

The prevalence and mean intensity of *A. oreini* in various fishes of Wular lake of Kashmir Valley during winter season (1988-89), shown by these authors is given in following table:

Name of the fish examined	No. of the fish	No. of fish infected	%age	Intensity & No. of worms
Schizsthorax niger	387	129	33.3	5.8 (753)
S. hugelii	182	65	35.7	5.8 (379)
S. esocinus	30	07	23.3	5.5 (39)
S. curvifrons	51	15	29.4	2.4 (37)
Oreinus plagiostomus	55	10	10.1	5.5 (55)
Oreinus sinuatus	78	23	29.4	6.3 (145)

Source: Dhar & Peerzada (1992).

The initial data presented by Dhar & Peerzada (1992) on the prevalence and percentage of fish infection covered a total of six species of fishes, infected by *A. oreini,* which is here being modified, keeping in view the recent nomenclatural changes pertaining to fish hosts, reducing the actual number of fishe species to 4 instead of six, which are, *viz. S. niger, S. esocinus, S. curvifrons* and *S. richardsonii*. The highest percentage of infection, therefore, was prevalent in *S. esocinus* (33.9%), followed by *S. niger* (33.3%), *S. curvifrons* (29.41%), as presented in the following modified table:

Name of fish	No. of fish	No. of fish infected	%age
Schizsthorax niger	387	129	33.3
S. esocinus (= *S. hugelii*)	30 182	07 65	33.96
Total	**212**	**72**	
S. curvifrons	51	15	29.41
S. richardsonii (= *Oreinus plagiostomus*) (= *O. sinuatus*)	55 78	10 23	
Total	**133**	**33**	**24.81**

Modified after Dhar & Peerzada (1992).

Majidah & Khan (1998) during the course of their investigation on seasonal population dynamics of two cestode parasite, *A. oreini* and *Gangesia* sp. in fishes such as *S. niger, S. esocinus (= S. hugelii), S. curvifrons, S. richardsonii (= O. plagiostomus), Triplophysa kashmirenis (= Nemacheilus kashmirensis)* and *Cyprinus carpio specularis* of Wular lake found that the infra population concentration was relatively more during summer. Further, the maximum infection was noticed in *S. esocinus, S. curvifrons* and *S. niger.*

Recently, Chishti & Ahmad (2003) reported the existence of cestode parasite *Adenoscolex oreini,* infecting the intestine of fishes, *viz. Carassius carassius* and *S. curvifrons* from Dal lake (urban eutrophic) and Manasbal lake (rural mesotrophic). Cestode species *(A. oreini)* was observed to be having two infection periods as winter and spring.

2.2.2.2 Amphibian and reptilian cestodes

It was Fotedar (1986), who for the first time reported the existence of nematotaeniid cestode, *Nematotaenia kashmirensis,* infecting small intestine of toad *(Bufo viridis)* in various regions of Kashmir Himalaya. The infection due to *N. kashmirensis* was reported to be very common among toads at many localities, *viz.* Srinagar, Baramulla, Sopore, Anantnag, Verinag besides, high-altitude areas (c 7,000–9,000 ft) like Pahalgam and Gulmarg. Heavy infection of *N. kashmirensis* was also prevalent in toads of these region and in some cases of infected amphibian, the distal region found bulged out due to the presence of large number of parasites, coiled together in luman, showing hardly any space for food to pass. The number of cestodes, in such cases was as high as twelve, of different length, and the incidence was found during August to September month.

Raina *et al.* (1973, 77) reported the occurrence of an Anocephalid Cestode parasite, *viz. Oochoristica tuberculata*, infecting the intestine of lizard, *Agama tuberculata* in Kashmir region besides, a description of teratological abbrration in *O. tuberculata*, collected from *A. tuberculata*, has also been provided by these authors.

2.2.2.3 Birds (wild and poultry) cestodes

In Kashmir region, so far four Species of cestode, infecting the intestine of *Corvus monedula* (Eastern Jackdaw) and *C. macrorhynchos*, (Himalayan jungle crow), have been reported. The parasites reported include *Anomotaenia galbulae, Choanotaenia infundibulum, C. microcantha* and *Mayhewia kavinia* (Fotedar & Chishti, 1974, 76; Chishti & Khan, 1984; Chishti *et al*., 1987).

Bird, *viz. Acridotheres tristis* of Kashmir, has been found to be infected by Cestode parasite *(Choanotaenia fotedari).* Castodes: *Anomotaenia acrocephali* and *Choenotaenia orioli*, are known to infect intestine of birds,*viz.*Great Red Warbler (*Acrocephalus stenotorecus brunnescens)* and *Oriolus oriolus* of Kashmir respectively (Fotedar & Chishti, 1974*a*, 76).

Indian Little Greb or Dabchick, *Podiceps ruficollis capensis* has been observed to be infected with intestinal cestode parasites, *viz. Pseudoschistotaenia indica* and *P. pindchii* by Fotedar & Chishti (1980) and Chishti (1982). In addition to this bird, Fotedar *et al.* (1970) and Chishti (1980), have observed the intestine of birds such as, Mallard or Wild Duck *(Anas platyrhynchos)* and Common Blue Rock Pigeon *(Columba livia neglecta)*, in Kashmir region, infected with parasite, *Dilepsis fotedari* and *Raillietina (R.. torquata* var. *kashmirensis* respectively.From Anchar Lake(Kashmir),Fotedar & Chisthi(1973) reported the existence of cestode-*Anomotaenia kashmirensis* in the intestine of bird, *viz. Sturnus vulgaris*

From above,it is clear that a total of 12 species of cestodes,distributed over 6 genera are known to infect about 13 species of wild birds (Table 2.8, pages 38-39).

Domestic/Poultry Birds

Earlier, Fotedar & Chishti (1974*a*) reported the existence of cestode parasite, *Choanotaenia infundibulani*, infecting the intestine of domestic fowl, *Gallus domesticus* of Kashmir region. The detailed studies on the occurrence and seasonal variation of Cestode parasite infection of domestic fowl (indeginous), *G. g. domesticus* in this valley, have been done by Fotedar & Khateeb (1987). Their survey has revealed a total of three Cestode parasite, *viz. Amoebotaenia sphenoides, Davainea proglottina* and *Raillietina tetragona.* The highest incidence of these parasites was shown by these authors in the month of September and lowest in the month of December and January besides, the prevalence and mean worm burden showed a decrease with decreasing temperature and rainfall.

Recently, Pandit *et al.* (1991) studied thoroughly the prevalence of helminth parasites in indeginous fowl of Kashmir, including various cestode parasites. The different intestinal (small) cestode parasites of fowl, with the percentage of prevalence were: *Raillietina tetragone*, 20%; *R. echinobothridia*, 10%; *Choanotaenia infundibulum*, 3%; *Hymenolepis carioea*, 2% and *Cotugnia diagonopora*, 1%.

The above authors also observed that the Cestodiasis infection in the fowl as compared to other helminth parasites was 45%. The infections were single, however, mixed infection with two to three species was not uncommon in Kashmir. Like other helminth parasites, cestode infection is responsible for causing much damage to poultry besides, they become susceptible to other diseases also. The occurrence of helminth diseases has been shown to be influenced by varying climatic and ecological factors.

Till-to-date date about 7 species of cestodes, distributed over 6 genera have been found in Kashmir,infecting domestic fowl(indeginous) (*Gallus domesticus*)(see Table 2.8, page 38).

2.2.3 Rat and domestic animal cestodes

Rat:

Raina *et al.* (1978) for the first time recorded the existence of cestode parasite, *viz. Mathevotaenia symmetria* from the intestine of common rat, *Rattus rattus* in Kashmir.

Cattle:

Eleven species of cestode parasites, infecting intestine and different organs/parts of ovines, sheep and goat, and bovids and yak of Kashmir, and Ladakh region so far recorded are: *Avitellina centripunctata, A. woodlandi, A. chalmersi, Stilesia globapunctata, Thysaniezia himalayai, Thysaniezia* sp.(Theysanosomatina: Anoplocephilidae); *Moniezia expansa, Moniezia* sp. (Anoplocephalinae: Anplocephalidae); *Coenurus* sp., *Taenia* sp. and *Echinococcus* sp. (Taeniidae) (Fotedar & Bambroo, 1978; Makdoomi *et al.*, 1995; Ranga Rao, 1994; Raina, 1973,75; Sudhan & Shahardar, 1999 and Tak *et al.*, 2003).

In Ladakh, Yak (*Bos grunniens)* is known to be infected with tapeworm, *Moniezia* sp., affecting intestine. In addition to this Cestode species, *Coenurus* sp. in the peritoneum in the form of cyst & *Echinococcus* sp. (cyst) in liver and lungs of yak, have also been reported from Ladakh by Ranga Rao *et al.* (1994).

Tak *et al.* (2003), have shown the wide prevalence of hydatid disease among bovids of Kashmir. The existence of Coenurus cyst in sub-cutaneous tissues of ear of goat, belonging to *Taenia* sp. (metacestode) for the first time, has been indicated by Sudhan & Shahardar (1999).

In Kashmir Valley, the incidence of cestode, *viz. Avitellina centripunctata, A. woodlandi, A. chalmarsi, Moniezia expansa, Stilesia globipunctata* and *Tysaniezia,* infecting sheep and goat in Kashmir, has been thoroughly studied by Raina (1973). He showed the incidence of *Avitellina* spp., *Moniezia* sp. and *Stilesia* sp., reaching to its maximum during March to August. The overall percentage infection of sheep and goat was 40.83%. These parasites were shown to have little variation in their percentage occurrence, almost equally common. However, *Avitellina* has been shown to possess maximum percentage of incidence, followed by *Stilesia* and lowest percentage being that of *Moniezia* sp.

The above author also investigated the weight-length relationship of three cestodes and correlated with worm burden. The average weight per cm. length was highest in *Moniezia* (14mg/cm) and lowest in *Stilesia* (3.5 mg/cm). The *Avitellina* was intermediate between the two cestode species as 5.1 mg/cm. The high weight per cm length in case of *Moniezia,* indicated a heavy drain on the host and hence lesser number of these worms per host, for a greater number could be lethal.

2.2.4 Hydatid cyst diseases of human and domestic animals of Kashmir valley

The hydatid disease, *viz.* hydatidosis (unicellular hydatid cyst), caused by cestode, *Echinococcus granulosus* (Taeniidae; Cyclophyllidae), is reported to be commonly occurring in domestic animals in many regions and localities of Kashmir. The hydatid diseases (larval tape worm type), *viz.* hydatidosis (unicellular hydatid cyst) and Alveococcosis or Echinococosis (multilocular or alveolar hydatid cyst), caused by *Echinococcus granulosus* and *E. multilocularis* respectively, belong to the family Taeniidae, order Cyclophyllidae, class Cestoda of Phylum Platyhelminthes. These hydatid diseases are not only serious parasitic infection but also affecting human and domestic animal populations since the time immorial. In Kashmir Valley, hydatid diseases are showing wide prevalence in human population, besides hydatidosis is also reported to be commonly occurring in domestic animals in many localities and regions of valley.

(a) Hydatidosis in Human and Animal (unicellular hydatid cyst diseases)

(*i*) ***In human population:*** In human population, *C. granulosus* is responsible for the formation of unicellular hydatid cyst, causing disease known as hydatidosis, measures from 2-8 mm in length and having single gravid proglottid, longer then wide and lodges typical taenia like eggs. The definite/natural or wild hosts of this cestode are dogs, wolves, foxes, etc. The parasite occurs in hundreds/thousands in these hosts and may be practically harmless. The infection of this disease reach to humans and domestic animals as their intermediate host through close association with dogs or through eggs from feces of dogs, wolves, or foxes, gaining entry into the intermediate hosts like humans, further it may be either through contaminated water or forage.

The development of hydatid cyst is slow. The young larva changes into hollow bladder, around which the host adds an enveloping, fibrous cyst wall. The fluid of the cyst is almost colourless, however, in older cysts in human tissues, is characterized by multiple daughter bladders or 'blood capsules', with multiple invaginated protoscoleces, budding from their walls (cystic cavity or inner layer of germinal epithelium) and the material consisting of dead and generated scoleces in milky fluid of parent cyst is referred to as hydatid sand.

In Jammu and Kashmir region, the hydatid cysts show wide diversity in different sites and organs of human body, include abdominal wall, brain, kidney, lungs, liver, neck, pancreas, peritoneum, ovary, sigmoid mesocolon, spleen, urinary bladder, and uterus. Besides these, multiple cysts in liver and lungs, liver and spleen, liver and peritoneum, have also been reported from Kashmir Valley.

The hydatid cysts cause much harm only when they grow to outrageous size and press on other organ/ s or when they liberate their fluid by leaking or rupturing, precipitating allergic symptoms. The formation of hydatid cysts in human and domestic animal represents the end of the life-line for the parasite.

The prevalence and epidemiological studies on human hydatid disease in populations in different areas of all the districts of Kashmir Province from 1998-99, have been made by Chishti *et al.*(2001). During this period, these authors found a total of 170 patients suffering from hydatid disease. The observations have been made in this connection, on patients suffering from this disease, and admitted in various surgical wards of Sher-i-Kashmir Institute of Medical Sciences (SKIMS). The district-wise percentage of infection of this disease was shown as: 26.47%, 21.76%, 18.82%, 11.88% and 5.38%, in district Anantnag, Pulwama, Baramulla, Srinagar, Budgam and Kupwara respectively.

The aforementioned authors have also given the percentage of cyst of organ site as: 52.94%, 35.29%, 7.05%, 2.3%, 1.17%, 0.53% and 0.53% in organ liver (isolated), lungs (isolated), liver and lungs, liver and spleen, liver and peritoneum, spleen, ovary respectively. Apart from this, their studies have shown the percentage of age-distribution and sex incidence of disease. The percentage is as: from 0-10 years 8.82%; 11-20 years, 31.76%; 21-30 years, 24.11%; 31-41 years, 18.82%; 41-50 years, 8.87%; 51-60 years, 5.29% and 60-70 years, 2.35%. The high incidence of disease was found in males (60%) and 40% of females were found to be positive for the disease. The susceptibility of males to infection than females, is mainly due to their (males) association with domestic animals during rearing.

The prevalence of human hydatid disease in Kashmir valley, in the recent years, have been updated by Fomda *et al.* (2002), based on clinically detected cases of disease in 5808 patients (4422 male and 1386 female), with age group ranging from 12 to 68 years, attending SKIMS, Soura, Kashmir, between April 1984 – April 2001. Further, the authors have shown the commonest site of infection as liver (60%), followed by lungs (33.23%), kidney (3.71%), ovary (1.18%), brain (1.01%), pancreas (0.34%), multiple cysts in liver and lungs (0.20%), spleen (0.12%), urinary bladder (0.08%) and abdominal wall (0.01%). The results of this study indicated a high prevalence of hydatid disease in the Valley and demands prompt control measures.

(*ii*) ***In animals (slaughtered) and Yak:*** The prevalence of hydatidosis in slaughtered domestic animals, *viz.* cattle, sheep, goat and buffaloes, at different abattoirs in Kashmir Valley, during August, 2001 to July, 2002, was investigated by Tak *et al.* (2003). The overall percentage infection rate of the disease was shown as 38.59%, 26.00%, 14.47% and 8.00 in cattle, sheep, goats and buffaloes respectively. The organ-wise distribution of hydatid cysts as shown by authors was that simultaneous involvement of both lung and liver, was more frequent in sheep (71.15%) and goat (54.54%). In cattle, lungs showed (63.63%), comparatively higher infection than liver. These studies also highlighted that right basal lungs were more infected than left ones. Spleen, kidney and heart of cow were only exceptional organs for the existence of hydatid cyst. In buffaloes, 100% infection was observed in lungs. Also, the authors showed that the fertility of hydatid cyst was highest in sheep (98.36%), followed by buffaloes 50%, goat (27.27%) and the least in cattle (9.09%).

Ranga Rao *et al.* (1994) during the course of their extensive investigations on parasitic infection of Indian Yak-*Bos (Poephagus) grunniens* in Ladakh region, have detected the occurrence of hydatid cyst in these domestic animals. They recovered majority of these cysts from liver (fertile) as well as lung (infertile) of these animals.

(b) Alveolar or multilocular hydatid disease (Echinococcosis) in human

The adult worms pertaining to multilocular or alveolar hydatid cysts of *E.multilocularis*, range from 3-5 mm long, are commonly found in liver of man. *E.multilocularis* causes disease in human known as polycystic echinococcosis. The definitive hosts of this parasite are: fox, cats, dogs and wolves, etc. The human cases of multilocular hydatid disease result when fruit and vegetable contaminated by faeces, containing eggs from infected canines are consumed by human or water contaminated by egg-containing faeces may be another source of infection. Apart from this, the intimation of man with dogs and cats can also result into human infection.

The growth and proliferation of *E.multilocularis* are extreamely slow. The larva of this parasite is sponge like, constantly growing mass of small separate vesicles embedded in a fibrous tissues. The vesicle contains a gelatinous substance instead of fluid. Roots grow out into neighbouring tissues. The central portion degenerates and dies, while growth continues on the outside. This results in a cluster of comparative smaller cysts that metastize through the surrounding tissue. In this type of cyst, there is neither protoscoleces nor hydatid sand and this type of hydatid cyst is known as multiloular or alveolar cysts, developing principally in liver. If the development of cyst reaches such a stage that they cannot be removed surgically, in that case death commonly results.

In Kashmir valley, alveolar hydatid disease has been reported from Uri region by Khuroo *et al.* (1980). The patient having multicellular hydatid cyst was also found to be suffering from Budd-Chari syndrome. This syndrome occurs due to thrombosis of hepatic veins and often the adjacent vena cava. Also, in this syndrome, the liver becomes tender, swollen and ascites, with portal hypertension and esophageal varices. This is usually a serious disease and obstruction can be removed surgically, to prevent death.

The diversity of hydatid cyst infection sites, in human population of Jammu & Kashmir region, reported by different authors, is given as under:

S. No. (a)	Infection site in human body locality (b)	Author(s) reporting (c)
	I. Unilocular hiydatid cys disease (Hydatidosis)	
	(a) Single cyst	
1.	Lungs, Rural areas of Kashmir Valley	Tak (2000) Fomda *et al.* (2002); Chishti *et al.*
		(2001)
2.	Sigmoid mesocolon, Kashmir	Ahmad *et al.* (2001)
3.	Brain, Rural Kashmir	Wani *et al.* (1999); Fomda *et al.* (2001)
4.	Neck-Cervicofacial region under sternomastoid muscle, Kashmir	Pampori *et al.* (2003)

Table contd.....

(a)	(b)	(c)
5.	Uterus (wall and vault), Kashmir	Fazili *et al.* (2002)
6.	Pancreas, Kashmir	Yattoo, *et al.* (1999); Misger *et al.* (2000); Fomda *et al.* (2002)
7.	Kidney, Jammu and Kashmir	Gupta *et al.* (2002); Fomda *et al.* (2002)
8.	Liver, Kashmir	Khuroo *et al.* (1991,99); Malik *et al.* (1997)
9.	Spleen, Kashmir	Chishti *et al.* (2001)
10.	Urinary bladder, Kashmir	Fomda *et al.* (2002)
11.	Abdominal wall, Kashmir	-do-
12.	Ovary, Kashmir	-do-
	(b) Multiple cysts	
13.	Liver and lungs, Kashmir	Chishti *et al.* (2001); Fomda *et al.* (2002)
14.	Liver and spleen, Kashmir	Chishti *et al.* (2001)
15.	Liver and peritoneum, Kashmir	-do-
	II. Multiple/Alveolar cyst Disease (Alveococcosis)	
16.	Alveolar hydatid disease of liver, with Budd-Chiari syndrome, Uri (Kashmir)	Khuroo *et al.* (1980).

2.2.5 Other human cestode diseases (hymenolepiasis and taeniasis)

Hymenolepiasis disease in human is caused by Cestode,*Hymenolepis nana*. *H. nana* belongs to the family Hymenolepididae and is also called as dwarf tapeworm of man, because of the fact that it is the smallest adult intestinal tapeworm found in man. The disease caused by this Cestode parasite is, characterized by malnutrition, diarrhoea and abdominal pain. In severe cases, there may be convulsions, epilepsy and insomnia.

H. nana shows a world-wide distribution, including India and J&K (Jammu). According to Faust *et al.* (1974), this parasite is more common in children than in adults and usually found in family, and Institutional group then in the population; more prevalent in warm than cold climate region.

In human hymenolepiasis there is no intermediate host and the infection is caused directly from patient to patient. *H. nana* is also parasitic on rodents. The young childrens are sometimes susceptible to infection from murine sources. In the hymenolepiasis, the infection is caused by ingestion of *H. nana* eggs from soiled linen, toilet seat or directly from anus to mouth (oro-faecal) and personal hygiene.

Arora *et al.* (1976), during the course of their study on the incidence of intestinal parasites in a rural community in Lamberi village of district Rajouri (Jammu), found *H. nana* infection in 1.8% of population examined and significantly higher prevalence rate was in the age group below 15 years (29%) than in the age group of 15 years and above (0.5%).

Taeniasis disease in human population in Kashmir region is reported to be caused by cestode, *viz. T. saginata* or beef tapeworm of man, belongs to Cyclophylldean group under family Taeniidae.

This cestode parasite is the common and large parasite, showing cosmopolitan distribution, infecting small intestine, by living of adult worms, with their heads embedded in the mucosa of the intestine.

The human infection due to *T. saginata* is acquired by eating raw/uncooked and 'measly' or insufficiently cooked beef/or meat of cattle, containing viable cysticercus larvae. Cattle become infection from grazing on ground, polluted through faecal matter of human, containing eggs of parasite. Cattle and allied animals act as intermediate host, and man acts as definitive host of *T. saginata.*

The common symptoms of this disease are abdominal pain excessive appetite, weakness and loss of weight. Due to large size of *T. saginata,* it is responsible for causing considerable disturbance in the normal functioning of digestive tract. At times, the worms may produce acute intestinal stoppage or sometimes the proglottids enter in the appendiceal luman and cause appendicitis. In taeniasis, usually the absorbed by-products cause systemic intoxication in the patients.

In Kashmir, Ahmad (1978) during the course of his investigation on the helminth parasites of rural population of Kashmir found 1.5% cases of *T. saginata.* The highest incidence was located in school-going age group (5 – 15 years). Koul *et al.*(2000) reported the prevalence of *T. saginata* in 185 patients (122 males and 63 females), having age from 3 – 60 years from this region.

TABLE 2.7. Systematic position and host-diversity of cestode parasite genera (Phylum: Platyhelminthes) of vertebrates (fishes, amphibians, reptile, birds, mammals, including human) in J & K state (India)

(I) Class: Cestoda; Sub-Class cestodaria				
Order	**Family**	**Sub-family**	**Cestode genus**	**Host/s**
Caryophyllidea	Caryophyllaeidae	Capingentinae	*Adenoscolex*	Fishes
(II) Class: Cestoda; Sub-Class cestodaria				
Order (a)	**Family (b)**	**Sub-family (c)**	**Cestode genus (d)**	**Host/s** -do-
Pseudophyllidea	Bothriocephalidae	-	*Bothriocephalus*	-do-
-do-	Diphyllobothriidae	-	*Ligula*	-do-
-do-	Proteocephalidae	Proteocephalinae	*Proteocephalus*	-do-
-do-	-do-	-do-	*Gangesia*	-do-
Cyclophyllidea	Anoplocephalidae	Linstowiinae	*Methevotaenia*	Rat
-do-	-do-	Thysanosomatinae	*Avitellina*	Sheep & Goat
-do-	-do-	-do-	*Stilesia*	-do-
-do-	-do-	-do-	*Thysaniezia*	-do-
-do-	-do-	Anoplocephalinae	*Moniezia*	-do-
-do-	-do-	-do-	*Oochoristica*	Lizard
-do-	Davaineidae	Davaineinae	*Cotugnia*	Fowl (domestic)
-do-	-do-	-do-	*Davainea*	—do—
-do-	-do-	-do-	*Raillietina*	Wild & domestic birds
-do-	Dilepididae	Dilepidinae	*Dilepsis*	-do-
-do-	-do-	-do-	*Amoebotaenia*	Domestic fowl
-do-	-do-	-do-	*Anomotaenia*	Wild bird
-do-	-do-	-do-	*Choanotaenia*	Wild & domestic birds
-do-	Hymenolepididae	Hymenolepidinae	*Hymenolepis*	Domestic fowl, Human
-do-	-do-	-do-	*Mayhewia*	Wild bird

Table contd....

(a)	(b)	(c)	(d)	(e)
-do-	Nematotaeniidae	-do-	*Nematotaenia*	Amphibian (toad)
-do-	Taeniidae	-do-	*Taenia*	Human
-do-	-do-	-do-	*Echinococcus*	Domestic animal, Human
-do-	-do-	-do-	*Coenurus (Cyst)*	Goat, Yak
-do-	Ambiliidae	-do-	*Pseudoschistotaenia*	Wild birds

Note: The cestode genera, listed in the above table, have been classified in the light of system of classification, given by Yamaguti (1959/85).

TABLE 2.8 Castode-fauna of vertebrates (fishes, amphibian, reptile, birds and mammals, including human) in Jammu, Kashmir & Ladakh region

Host group/Cestode species (a)	Host species (b)	Location of infection (c)	Locality (d)
1. **FISHES**	*Adenoscolex oreini richardsonii* (= *Oreinus sinuatus*), *O. plagiostomus*), *S. esocinus* (= *S.hugelii*), *S. curvifrons*, *S. niger*, *Triplophysa kashmirensis* (Syn. *Nemocheilus kashmirensis*), *Carassius carassius* and *Cyprinus carpio specularis*	Intestine	Dal Lake, Manasbal Lake and various other waterbodies of Kashmir region
2. *Adenoscolex* sp.	*Schizothorax* sp.	-do-	Anchar Manasbal & Wular lake of Kashmir.
3. *Bothriocephalus - acheilognathi*	*Cyprinus carpio communis*, *Schizothorax* sp.	-do-	River Jehlum (Chattabal) Srinagar.
4. *Gangesia (G.) fotedari*	*Glyptothorax*	-do-	Wular Lake
5. *Gangesia (G.) lucknowi*	-do-	-do-	Baramulla (Kmr.)

Table contd....

(a)	(b)	(c)	(d)	(e)
6. *Gangesia (Gangesia)* sp.	*S. niger, S. esocinus (=S.hugelii)S.curvifrons, S. richardsoni (= Oreinus plagiostomus), Triplophysa kashmirensis (=N. kashmirensis), C.c. specularis, Glyptothorax* sp.		-do-	Kashmir
7. *Gangesia (Vermia) jummuensis*	*Wallago attu*		-do-	Jammu
8. *G.(v.) kashmirensis*	*Glyptothorax*		-do-	Baramulla (Kashmir)
9. *Ligula intestinalis*	*Schizothorax* sp.		-do-	Dal Lake, Srinagar.
10. *Protocephalus kashmirensis*	*Triplophysa kashmirensis (=Nemacheilus kashmirensis)*		-do-	Kashmir
11. *Protocephalus torulosus*	-do-		-do-	Harvan, Kashmir
II. AMPHIBIAN (TOAD)				
1. *Nematotaenia kashmirensis*	*Bufo viridis*		Intestine	Srinagar, Baramulla, Sopore, Anantnag, Verinag, Pahalgam and Gulmarg
III. REPTILE (LIZARD)				
1. *Oochoristica tuberculata*	*Agama tuberculata*		-do-	Kashmir
IV. BIRDS (WILD & DOMESTIC /POULTRY)				
1. *Amboebotaenia sphenoides*	*Domestic fowl*		-do-	-do-
2. *Anomotaenia acrocephali*	*Acrocephalus stentoreus brunnescens (Great Red Warbler)*		-do-	-do-
3. *Anomotaenia galbulae*	*Corvus monedula, C. macrorhyrnchos*		-do-	-do-
4. *Anomotaenia kashmirensis*	*Sturnus vulgaris*		-do-	Anchar, Srinagar.
5. *Choanotaenia fotedari*	*Acridotheres tristis*		-do-	Kashmir
6. *Choanotaenia infundibulani*	*C. splendens, Gallus domesticus,* domestic fowl (indigenous)		-do-	-do-

Table contd....

(a)	(b)	(c)	(d)
7. *Choanotaenia micacantha*	*Corvus monedula*	-do-	-do-
8. *Choanotaenia orioli*	*Orioleis oriolus*	-do-	-do-
9. *Cotugnia diagonopora*	Domestic fowl (indigenous)	-do-	-do-
10. *Davainea proglottina*	Domestic fowl (indigenous)	-do-	-do-
11. *Diplesis fotedari*	*Anas platyrhynchos*	-do-	-do-
12. *Hymenolepis carioea*	Domestic fowl (indigenous)	-do-	-do-
13. *Mayhewia kavini*	*Corvus monedula*	-do-	-do-
14. *Psendoschistotaenia indica*	*Podiceps ruficollis capensis*	-do-	-do-
15. *Psendoschistotaenia pindchii*	*Podiceps ruficollis capensis*	Intestine	Kashmir
16. *Raillietina echinobothridia*	Domestic fowl	-do-	-do-
17. *Raillietina tetragona*	-do-	-do-	-do-
18. *Raillietina torquate kashmirensis*	Blue Rock Pigeon	-do-	-do-
(V) MAMMALS (WILD, DOMESTICATED) & HUMAN			
(A) RAT			
1. *Methevotaenia sysmmetrica*	*Rattus rattus*	Intestine	Kashmir
(B) DOMESTIC ANIMALS (SHEEP, GOAT, BOVIDS & YAK)			
1. *Avitellina centripunctata*	Sheep & Goat	Intestine	-do-
2. *Avitellina* (A.) *woodlandi*	-do-	-do-	-do-
3. *Avitellina chalmersi*	-do-		
4. *Coenurus* (Cyst)	Yak	Peritoneum	Kashmir & Ladakh
5. *Echinococcus* (Hydatid disease)	Ovines, Bvids and Yak	Liver, lung, spleen, kidney and heart	-do-

Table contd....

(a)	(b)	(c)	(d)
7. *Moniezia* sp.	Ovines and Yak	-do-	Kashmir & Ladakh
8. *Stilesia globipunctata*	Sheep & Goat	-do-	Kashmir
9. *Taenia* sp.	Goat	Ear (sub-cutaneous tissues)	-do-
10. *Thysaniezia himalayai*	Sheep (local)	Intestine	-do-
11. *Thysaniezia* sp.	Sheep & Goat	-do-	-do-
C) HUMAN			
1. *Echinococcus granulosus*	Human	Cysts in liver, lungs, brain, neck uterus, pancreas, kidney, spleen, urinary bladder, abdominal wall and ovary.	Jammu and Kashmir region
2. *Echinococcus multilocularis*	-do-	Alveolar hydatid cyst of tiver.	Uri, Kashmir
3. *Echinococcus* sp.	-do-	Pancreatic head hydatid cyst	Kashmir
4. *Hymenolepsis nana*	-do-	Intestine	Rajouri, Jammu
5. *Taenia saginata*	-do-	—do—	Kashmir

Sources: (Table 2.8, pp 37-40):

(i) Fish Cestodes: Ahmad & Chishti (1998); Ara *et al.* (2000); Chishti & Ahmad (2002); Datta (1936); Dhar & Fotedar (1980); Dhar & Majidah (1983, 84, 87); Dhar & Peerzada (1992); Dhar & Raina (1983); Fotedar (1958); Gupta (1967); Majidah & Khan(1996).

(ii) Amphibian (toad) Cestode: Fotedar (1966)

(iii) Reptilian (lizard) Cestode: Raina *et al.* (1973, 77)

(iv) Birds Cestodes:

(a) Domestic/Poultry birds: Chishti (1980); Fotedar & Chishti (1974); Fotedar & Khateeb (1987) and Pandit *et al.* (1991).

(b) Wild birds: Chishti (1973, 82); Chishti *et al.* (1987); Chishti & Khan (1984); Fotedar & Chishti (1973, 74, 76 & 80); Fotedar *et al.* (1970).

(v) Mammals:

(a) Rat: Raina *et al.* (1978)

(b) Domestic mammals (sheep, goat, bovids and yak): Fotedar & Bambroo (1978), Makdoomi *et al.* (1995). Ranga-Rao (1994); Raina (1973,75); Sudhan & Shahardar (1999) and Tak *et al.* (2003)

(c) Human: Ahmad (1978); Ahmad *et al.* (2001); Arora *et al.*; (1976); Chishti *et al.* (2001); Fazili *et al.* (2003); Fomda *et al.* (2002); Pampori *et al.* (2003); Koul *et al.* (2000); Khuroo *et al.* (1980, 91, 93); Tak (2000).

CHAPTER 3

ACANTHOCEPHALAN PARASITES

3.1. GENERAL ACCOUNT

Acanthocephalan parasites are commonly known as spiny-headed worms. The adults are endoparasites of various vertebrates, living in the alimentary tract, characterized by the presence of protrusible armed proboscis at the anterior terminal. The sexes are separate. The females are longer in size than males. van Cleave (1948), Hymen (1951) and Cheng (1973) gave Acanthocephalan a status of Phylum, Cheng (1973) included three orders under this Phylum as: Archiacanthocelphala (parasites of terrestrial hosts); Palaeacanthocephala (parasites of fishes, reptiles birds, some terrestrial reptiles and mammals and invertebrate crustacean) and Eoucanthocephala (parasites of fishes, reptiles and crustacean). Petrotschenko (1956) regarded Acanthocephala as a class and divided it into three sub-classes, *viz.* Neoechinorhynchinae, Echinorhynchinae and Giganthorhynchinae. Yamaguti (1963) also regarded Acanthocephala as a class however, included under it four orders as: Apororhynchidea, Neoechinorhynchidea, Echinorhynchidea and Giganthorhynchidea. For the classification and nomenclature of acanthocephalan parasite species, reported from various vertebrate hosts (fishes, amphibians and birds) so far from Jammu, Kashmir and Ladakh region, Yamaguti (1963) has been followed, and these species belong to orders *viz.* Neoechinorhynchidea and Echinorhynchidea.

The acanthocephalans are injurious to the host by damaging the tissues of the intestinal wall by embedding the hooks present on the proboscis. There is inflammation at the site of attachment when the worms move and reattach, the old sores become infected by bacteria. In heavy infection, there is loss of appetite and interference with digestion, leading the unthriftness. Fotedar & Dhar (1977) reported heavy infection with about 200 *Pomphorhynchus* worms in the intestine of single fish, *viz. Schizothorax richardsoni (=Oreinus plagistomus)* from River Tawi of Jammu. Hyman (1951), reported massive infection from other region of the world. Occasionally worms cause perforation of the gut wall and leading to a fatal peritonitis (Chandler & Read, 1961). A single worm species can exist in more than one species of host, as has been reported from Kashmir fishes. For example, Acanthocephalan, *viz. Pomphorhynchus kashmirensis* has been found to infect as many as eight (8) different species of fishes in Kashmir. The same acanthocephalan species has been found in different families of the fishes.

The life-cycle of Acanthocephalan requires an intermediate host, which is usually an arthropod (crustacean), for parasite Species of aquatic vertebrate hosts and grubs, roaches etc, for those of land animals. The eggs of acanthocephalans after passing out from vertebrate hosts are ingested by invertebrate hosts (arthropod). In this host, egg develops into acanthor stage. In host's gut, it penetrates the gut wall and move into haemocoel and thereafter transformed into a stage known as acanthella. After further development, the acanthella changes to juvenile known as cystocanth. This development in intermediate host takes from 6-12 weeks. When the arthropod host, harbouring with cystocanth is ingested by the appropriate vertebrate host, the cystocanth in the vertebrate intestine loses its sheath and develops into adult, with sexual maturity. In some cases, a second intermediate host is required.

3.2. SYSTEMATIC SURVEY AND DIVERSITY OF ACANTHOCEPHALANS

It was Datta (1936, 37), who for the first time studied the Acanthocephalans of Kashmir fishes. Various new species of Acanthocephalans described by him include *Acanthocephalus kashmirensis, Neoechinorhynchus devdevi (=Eosentis devedi), N. yalei (E.yalei).* Besides these, he also recorded *N. rigidus* and *N. rutili* from Kashmir Himalayan regions.

Detiled investigations of Acanthocephala of fishes and Amphibians of Kashmir have been done by Kaw (1941, 51). He described various new species of these worms from fish hosts, which are, *viz. Echinorhynchus orientalis, Neoechinorhynchus manasbalensis* and *Pomphorhynchus kashmirensis.* Apart from these, Kaw (1951) also reported *Pomphorhynchus sp.* (encysted young form) from fish *(Botia birdi).*

The first checklist of acanthocephalan parasites of vertebrates (fishes and amphibians) of Kashmir was provided by Gupta (1967). He listed about nine species of acanthocephala, infecting eight different species of fishes besides, two species parasitising frog *(Rana cyanophlyctis).* This checklist was based on the investigations carried out by earlier workers on these parasites from different regions and areas of Jammu, Kashmir and Ladakh.

In the recent years, various authors who described more new acanthocephalans from different fish species and amphibian (toad) are: Fotedar (1968) on *Neoechinorhynchus oreini* from *Schizothorax richardsoni (=Oreinus sinuatus);* Fotedar *et al.* (1970) on *Pomphorhynchus bufonis* from *Bufo viridis* (toad); Fotedar & Dhar (1977) on *Pomphorhynchus megacanthus* from *Cyprinus carpio spicularis,* on *Pomphorhynchus oreini* from *S. richardsoni (=O.sinuatus),* on *P. orientalis* from *Schizothorax niger* and on *P. tori* from *tor tor;* Dhar & Kharoo (1985) on *Neoechinorhynchus glyptosternum* from *Glyptosternum* sp.; Chishti & Bakshi (1992) on *Pomphorhynchus fotedari* from *Cyprinus carpio communis.*

Descriptions of two new acanthocephalans parasitizing the fishes of Jammu have been provided by Fotedar & Dhar (1977) on *Pomphorhynchus jammuensis* from *Schiothorax richardsoni (=Oreinus plagiostomus)* Koul *et al.* (1992) on *Pallisentis jagni* from *Channa channa.*

For the first time Fotedar & Dhar (1973) recovered a large number of Juvenile form of Acanthocephla, *viz. Polymorphus* sp. from frog *(R. cyanophlyctis).* These authors also found large number of Juvenile form of *Polymorphus* at different stages of development in the body cavity of Amphipod *(Gammarus* sp.), occurring in Murran area of Kashmir. They also successfully completed the Life-cycle studies of this worm by artificially infecting young ducks. Late Juveniles and adults were identified as *Polymorphus magnus.* These authors presume that present infection might be accidental, since the infected frog also inhabited the same water body in which the amphipods in large number were found to be infected.

The acanthocephalan species diversity in fish hosts of Jammu, Kashmir and Ladakh, in rivers and other water-bodies of these regions, showed higher number *i.e.* 5 spp. in *S. richardsoni,* followed by 4 each in *S. niger* and *Triplophysa kashmirensis,* 3 in *S. esocinus,* 2 each infecting *S. curvifrons, S. plagiostomus* and *Schinothorax* sp. Rest of the 10 Species of fishes showed the incidence of single acanthocephalan species infection, in each fish host species The acanthocephalan species diversity in fish hosts of Jammu, Kashmir and Ladakh,. In addition to these, Frog *(Rana cyanophlyctis)* showed the incidence of 3 parasite species and toad was reported to be parasitized by single Species. Also, invertebrate (Amphipod) and Duck were found to be infected with *Polymorphus* sp. and *Polymorphus magnus* respectively, in Murran region of Kashmir valley.

3.3 ACANTHOCEPHALANS OF FISHES OF WULAR LAKE

Datta (1936) was the first researcher to investigate the acanthocephalans, parasitizing the fishes of Wular Lake in Kashmir. Later, a detailed survey of acanthocephans of fishes of Wular Lake, for the first time, was conducted by Dhar & Majidah (1987). They reported the wide existence of two acanthocephalans, *viz. Neoechinorphnchus manasbalensis* from the intestine of fishes, *viz. S. esocinus, S. niger, Triplophysa kashmirensis (=Nemachilus kashmirensis)* and *Pomphorhynchus kashmirensis* from the intestine of *Schizothorax* and *Nemachiclus* sp.

Recently, the thorough investigations on seasonal population dynamics of *N. manasbalensis, P. kashmirensis* and *Pomphorhynchus* sp., parasitizing the fishes of Wular lake, have been conducted by Dhar & Peerzada (1994), Chishti & Peerzada (1998) and Majidah & Khan (1998).

Detailed findings on host spectrum and seasonal dynamics of *P. kashmirensis* and *N. manasblensis* by Chishti & Peerzada (1998), have revealed that of the six different fish species of Wular Lake, with both the parasites, were only prevalent in fishes, *viz. S. curvifrons, S. niger* and *S. richardsoni. Cyprinus carpio* and *S. esocinus* were only parasitized by *P. kashmirensis*. However, *Crossochiclus latius* did not show the prevalence of either of the acanthocephan (Table 3.1). These authors have also shown that the maximum prevalence of infection due to *P. kashmirensis,* was spring season in all fish hosts. Further, it was reported that the recruitment of infection was dominated by the presence of juveniles and post juveniles, and also males. The distribution of parasites was over dispersed in all the hosts so in length group of *Schizothorax niger*. An increase in the mean number of parasites per host with an increase in host length was also indicated by Chishti & Peerzada (1998).

TABLE 3.1: Host spectrum of Pomphorhynchus kashmirensis and Neoechinorhynchus manasbalensis in Wular lake (Kashmir)

S.No. (a)	Host Fish Species (b)	*P. kashmirensis* (c)	*N. manasbalensis* (d)
1.	*Cyprinus carpio*	+	-
2.	*Crossochilus latius*	-	-
3.	*Schizothorax curvifrons* *(= Schizothorax hugelii)*	+ +	+ +
4.	*Schizothorax esocinus*	+	-
5.	*Schizothorax niger*	+	+
6.	*Schizothorax richardsoni* *(=Oreinus plagiostomus)* *(= Oreinus sinuatus)*	+	+
7.	*Hybirds (Schizothorax* and *Oreinus)*	-	-

+ = Present in the fish host intestine, - = Absent in the fish host intestine

Source: Modified after Chishti & Peerzada (1998).

TABLE 3.2 *Fish Host spectrum of Acanthocephalans of Wular Lake of Kashmir Valley*

S.No.	Fish species	Acanthocephalan species
1.	*Cyprinus carpio*	PK
2.	*Nemachielus vittatus*	NR
3.	*Salmo trutta fario*	PK
4.	*Schizothorax carvifrons*	NM, PK
5.	*Schizothorax esocinus*	NM, PK
6.	*Schizothorax niger*	NM, PK
7.	*Schizothorax plagiostomus*	NM, PK
8.	*Schizothorax richardsoni*	NM, PK
9.	*Schizothorax* Species	PK
10.	*Triplophysa kashmirensis*	NM, NR, PK

NM = *Neoechinorhynchus manasbalensis.*

NR = *Neoechinorhynchus rutili*

PK = *Pomphorhynchus kashmirensis*

Through the scientific literature published so far on the acanthocephalans, parasitizing the fish species existing in Wular lake of Kashmir, it is evident that three acanthocephans, *viz. Neoechinorhynchus manasbalensis, N. rutili* and *Pomphorhynchus kashmirensis*. Of these worms, *P. kashmirensis* has shown maximum number of fish host species *i.e.* 9, followed by *N. manasbalensis* as 6 (six) and *N. rutili,* with 2 (two) fish host species only.

In conclusion, it may be stated that a total of 24 spp. of acanthocephan parasites belonging to orders, *viz.* Neoechinorhynchidea (family Neoechinorhynchidae, Quadrigyridae) and Echinorhynchidea (family Echinorhynchidae, Polymorphidae and Pomphorhynchidae), parasitizing vertebrates (fishes, amphibians and duck) and invertebrate (crustacean). About 19 spp. of acanthocephalans, infecting 17 different fish Species, belonging to different families, 3 spp. parasitizing frog *(Rana cyanophlyctis)*, one of the species also found to parasitize *Gammarus* (Amphipod: Crustacean), and one Species each infecting toad *(Bufo viridis)* and duck (bird), existing in different water-bodies of regions/localities of Jammu, Kashmir and Ladakh. (see table 3.3, pp. 47-50).

TABLE 3.3 – Acanthocephalan parasites of Vertebrates (fishes, amphibians and birds) of J&K State (India).

S. No.	Acanthocephalan & Host animal species (a)	Kind of host (b)	Location of infection (c)	Locality (d)	Authors (References) (e)
1.	***Acanthocephalus kashmirensis*** Datta	Fish	Intestine	Kashmir, Ladakh	Datta (1936)
	Schizopygopsis stoliczke Steindachner				
2.	***Echinorhynchus orientalis*** Kaw	-do-	-do-	Dal Lake, Kashmir	Kaw (1951)
	Schizothorax sp.				
3.	***Neoechinorhynchus cyanophlyctis*** Kaw	Frog	-do-	Srinagar, Kashmir	Kaw (1951)
	Rana cyanoplychtis				
4.	***Neoechinorhynchus devdevi*** (Datta)	Fish	-do-	Dal Lake,	Datta (1936)
	(= *Eosentis devdevi* Datta)			Srinagar	Tripathi (1956)
	Schizothorax niger (Heckel)			Manasbal	
	(= *Schizothorax planifrons* Heckel)				
5.	***Neoechinorhynchus glyptosternumi-*** Dhar & Kharoo *Glyptosternum* sp.	-do-	-do-	Kashmir	Dhar & Kharroo (1985)
6.	***Neocehinorhynchus hutchinsoni*** Datta	-do-	-do-	Leh, Ladakh	*cf* Tripathi(1956)
	Diptychus maculatus Steindachner				
	Triplophysa kashmirensis Hora	-do-	-do-	Tellbal Nallah, Srinagar	Raina & Koul (1983)
	(= *Nemachielus kashmirensis* Hora)				
7.	***Neoechinorhynchus manasbalensis*** Kaw	-do-	-do-	Wular	Cheshti & Peerzada
	Schizothorax curvifrons Heckel	-do-	-do-	-do-	(1998)
	(= *Schizothorax hugelii* Heckel)	-do-	-do-	-do-	-do-
	Schizothorax esocinus Heckel	-do-	-do-	-do-	Dhar & Majidah (1987)
	Schizothorax niger Heckel				Chishti & Peerzada
	Schizothorax plagiostomus Heckel				(1998) -do-
	Schizothorax richardsoni (Gray)	-do-	-do-	Wular	Chishti & Peerzada
	(= *Oreinus plagiostomus* Heckel)	-do-	-do-	Manasbal	(1918)
	(= *Oreinus sinuatus* Heckel)	-do-	-do-		Kaw (1951)Dhar &-

Table contd...

	(a)	(b)	(c)	(d)	(e)
	Triplophysa kashmirensis Hora (= *Nemachielus kashmirensis* Hora)	-do-	-do-	Wualr	Majidah (1987)
8.	***Neoechinorhynchus oreini*** Fotedar *Schinothorax richardsoni* (Gray) (= *Oreinus sinuatus* Heckel)	-do-	-do-	Anantnag	Fotedar (1987)
9.	***Neoechinorhynchus rigidus*** (Datta) (= *Eosentis rigidus* van Cleave) *Schizothorax zarudnyi*	-do-	-do-	Saistan, Kashmir	Datta (1937) Tripathi (1956)
10.	***Neoechinorhynchus rutili*** Muller *Triplophysa kashmirensis* Hora (= *Nemachielus stolliczkae* Steindaehna *Nemachieclus vittatus*	-do- -do-	-do- -do-	Wualr Lake -do-	Datta (1936) Kaw (1951)
11.	***Neoechinorhynchus yalei*** (Datta) (= *Eosentis yalei* Datta) *Schinothorax esocinus* Heckel	-do-	-do-	Jammu Manasbal	
12.	***Pallisentis jagni*** Koul, Raina, Bambroo & Koul *Channa channa*	-do-	-do-	Jammu	Koul *et al.* (1992)
13.	***Polymorphus magnus*** Skrjabin	Duck	Intestine	Murran, Kashmir	Fotedar & Dhar (1973)
14.	***Polymorphus sp.*** *Rana cyanophylctis* *Gammarus (Amphipod)*	Frog Crustacean	Rectum Body cavity	Murran, Kashmir Murran, Kashmir	Fotedar & Dhar (1973) Fotedar & Dhar (1973)
15.	***Pomphorhynchus bufonis*** Fotedar, Duda & Raina *Bufo viridis*	Toad	Intestine	Kashmir	Fotedar et al. (1970)
16.	***Pomphorhynchus dubium*** Kaw (sic dubious) (A juvenile form) *Rana cyanophlyctis*	Frog	Sub-perit- oneal region	Srinagar	Kaw (1941) Yamaguti (1963)

Table contd...

	(a)	(b)	(c)	(d)	(e)
17.	***Pomphorhynchus fotedari*** Chishti & Bakshi *Cyprinus carpio communis* Linnaeus	Fish	Stomach	Hokersar, Kashmir	Chishti & Bakshi (1992)
18.	***Pomphorhynchus jammuensis*** Fotedar & Dhar, *Schizothorax richardsoni* (Gray) (= *Oreinus plagiostomus* Heckel)	-do-	Intestine	Tavi River, Jammu	Fotedar & Dhar (1977)
19.	***Pomphorhynchus kashmirensis*** Kaw *Cyrinus carpio* Linnaeus	-do-	-do-	Wular	Chishti & Peerzada (1998)
	Salmo trutta fario Linnaeus *Schizothorax curvifrons* Heckel	-do-	-do-	Telbal, Harvan, Kashmir	Fotedar & Dhar (1977)
	(=*Schizothorax hugelli* Heckel)	-do-	-do-	Wular	Chisthi & Peerzada (1998)
	Schizothorax esocinus Heckel	-do-	-do-	Wular	-do-
	Schizothorax niger (Heckel)	-do-	-do-	Wular	-do-
	Schizothorax pleagiostomus Heckel (= *Oreinus sinuatus* Heckel)	-do-	-do-	Wular	Dhar & Majidah (1987)
	Schizothorax richardsoni (Gray) (= *Oreinus flegistomus* Heckel)	Fish	Intestine	Wular	Chishti & Peerzada (1998)
	Triplophysa kashmirensis Hora (= *Nemachielus kashmirensis* Hora)	-do-	-do-	Wular	Kaw (1941), Dhar & Majidah (1987)
20.	***Pomphorhynchus megacanthus*** Fotedar & Dhar *Cyprinus carpio spicularis* Lacepede	-do-	-do-	Telbal, Srinagar	Fotedar & Dhar (1977)
21.	***Pomphorhynchus oreini*** Fotedar & Dhar *Schizothorax richardsoni* (Gray) (= *Oreinus sinuatus* Hecked)	-do-	-do-	Ganderbal	Fotedar & Dhar (1972)
22.	***Pomphorhynchus orientalis*** Fotedar & Dhar *Schizothorax niger* Heckel	-do-	-do-	Ganderbal	Fotedar & Dhar (1972)

Table contd...

(a)	(b)	(c)	(d)	(e)	
23.	***Pomphorhynchus tori*** Fotedar & Dhar Dhar, *Tort tor* (Hamilton- Buchanan)	-do-	-do-		River Jehlum Fotedar & Dhar (1972)
24.	***Pomphorhynchus* sp.** (Encysted young form) *Botia birdi* Chaudhuri	-do-	Intestinal wall		Dal Lake, Kaw (1951) Srinagar

Systematic position of Acanthocephala: Serial No./s in the above table	**Order**	**Family**	**Sub-family**
3, 4, 5, 6, 7, 8, 9, 10,11	Neoechinorhynchidea	Neoechinorhynchidae	Neoechinorhynchinae
12	-do-	Quadrigyridae	–
1, 2	Echinorhynchidea	Echinorhynchidae	Echinorhynchinae
13, 14	-do-	Polymorphidae	Polymorphinae
15, 16, 17, 18, 19, 20, 21, 22, 23, 24	-do-	Pomphorhynchidae	–

Note: 1.Synonyms in case of hosts and acanthocephalan species are given in parentheses, below the species.
2. For nomenclature of Acanthocephala, Yamaguti (1963) has been consulted and, for nomenclature of fishes, Kullandar *et al.* (1999) and Talwar & Jhingran (1991), have been followed.

CHAPTER 4

NEMATODES (ANIMAL AND PLANT-PARASITIC) AND PARASITIC-LEECHES

4.1 ANIMAL NEMATODES

In J & K State, about 83 species of nematode parasites, distributed over 48 genera are known to infect not only vertebrates (fishes, amphibians, reptiles, birds and mammals, including human) but also invert-tebrates, like insects. The adult parasites range from microscopic to some centimeters. Recent authors like D. Hondt (1997) and Anderson (2000), have given nematodes a status of Phylum.

4.1.1 General account, species richness, systematic position and host diversity

The animal nematodes are elongated and cylindrical, tappering at both ends. Nematodes, *viz. Capillaria* spp. are uniform, cylindrical and extremely thin. Very often male nematodes are smaller than females. The nematodes are covered outside by very tough and impermeable transparent or semitransparent cuticle. This cuticle is usually marked from outside by markings, bristles, spines, ridges or expansions of various kinds.

The nematode Species in J & K State, have been found to infect various sites /systems of animals (wild and domesticated) as well as human beings, in various localities and areas. These include eye, stomach, intestinal tract, respiratory tract and body cavity, etc. All nematodes pass through four moults during their development and fourth stage larva has been assigned to specific forms of nematodes.

Vertebrates acquire their nematode parasites in different ways: i) the penetration of infective third stage larvae into skin of hosts (*Rhabdias* spp.); ii) through the ingestion of eggs containing infective larva (Oxyuroidea); iii) through corpophagy (e.g. pinworms); iv) through the ingestion of larvae in the tissues of intermediate and paratenic hosts (Camallanoids); v) through autoinfection (Cosmocercoids); vi) transplacental and transmammary transmission (Ascaridoids).

The nematode Species, infecting different host Species in Jammu, Kashmir and Ladakh regions, are high-lighted under the following heads:-

4.1.2 Insect, fish, amphibian, reptilian and birds (wild and poultry) nematodes

4.1.2.1 Insect nematodes

Earlier records of nematode parasites of insects of Kashmir, have been given by Basir (1941). He reported two new nematode species, *viz. Galebiella galebiella* and *G. islamabadi* from the intestine of aquatic beetle, inhabiting hill streams (c, 6,000 ft above sea level) in Anantnag region of Kashmir Valley. Gupta (1967), also included these two Species of insect nematodes in his paper on helminthic-fauna of Kashmir.

Description of a new species of nematode; *viz. Pseudonymus multiannulata*; infecting aquatic beetle (*Hydrophilus* sp.) in Srinagar, has been provided by Fotedar (1964). So far only three Species of nematodes, distributed over two genera are reported from Kashmir region, (table 4.3, p. 64)

4.1.2.2 Fish nematodes

The previous systematic and taxonomic studies on fish-nematodes of J & K State, have been done by: Chakravorti (1942), Thapar (1950), Gupta (1967), Fotedar & Dhar (1970), Raina & Dhar (1972), Fotedar & Dhar (1975), Dhar & Fotedar (1979), Dhar & Anwari (1982) and Chishti & Bakhshi (1990).

A total of nine Species of nematodes, belonging to four genera, under 3 different families (Camallanidae, Rhaldochoidae and Cystinicolidae), of order Spirurida and class Secernetea, has so far been recorded from this region of India. These nematodes have been found to infect intestine of eleven Species of fishes, endemic as well as exotic of Jammu and Kashmir regions (Table 4.2).

The studies conducted outside India, on development and transmission of nematode parasite infection of fishes, have established that fish acquire cystidicolid infection by ingesting arthopods (Ephemeoptera or crustaceans), containing third stage larva.The Camallanids usually involve copepods (crustaceans) as intermediate host and mayflies (insects) transmit infection in fishes due to Rhabdochonid parasites (Anderson, 2000).

The various nematode Species of different fish hosts include *Camallanus cotti (=Camallanus fotedari), C.. oxycephalus, C. solmonae, Procamallanus kashmirensis, Rhabdochona guptai, R. kashmirensis, R.*

Fish species	Nematode parasite species
1. *Cyprinus carpio specularis*	*C. cotti (= C. fotedari), R, guptai, R. kashmirensis, R. himalyai*
2. *Triplophysa kashmirensis (= Nemacheilus kashmirensis)*	*C. cotti, R. kashmirensis* and *R. himalayai*
3. *Nemacheilus sp.*	*R.. himalayai*
4. *Schizothorax richardsonii (= Oreinus plagiostomus) (= Oreinus richardsonii) (= Oreinus sinuatus)*	*C. cotti, R. kashmirensis, R. himalayai, Spinitectus sp., R. guptai* and *Camallanus oxycephalus, C. cotti, R. guptai, R. kashmirensis and R. himalayai*
5. *Schizothorax curvifrons*	*C. cotti. R, himalayai, R, kashmirensis*
6. *Schizothorax esocinus (= Schizothorex hugelii*	*C. cotti. R, himalayai R, kashmirensis*
7. *Schizothorax niger*	*C. cotti R, himalayai R, kashmirensis*
8. *Schizothorax sp.*	*C. oxycephalus, R, himalayai*
9. *Wallago attu*	*P. kashmirensis, R. kashmirensis*
10. *Salmo trutto fario*	*Rhabdochona sp.*
11. *Salmo sp.*	*C. salmoe*

Note: The synonyms of fish species in above cited table, given in the parentheses, are in accordance with the nomenclature given by: Talwar & Jhingran (1991) and Kullander *et al.* (1999).

himalayai, Rhaldochona sp. and *Spinitectus* sp. The host-fish spectrum of nematode parasites, in the light of recent revised fish species nomenclature, is given on page 52. From this table, it is evident that the highest number, i.e. 6, of different Species of nematode, have been reported in the fish, *viz. Schizothorax richardsonii,* followed by 4 spp. each in *Cyprinus carpio specularis* and *S. curvifrons,* 3 spp. each in *Triplophysa kashmirensis, S. esocinus* and *S. niger,* 2 spp. each in *schizothorax* sp. and *Wallago attu.* The rest of the fish host Species have been found to be infected with only Species of nematode parasite.

Detailed investigation, for the first time on fish parasitization by different nematode species, *viz. C. cotti (= C. fotedari)* from the intestine of *Schizothorax niger* and *R. himalayai* also from the intestine of *Schizothorax* sp. and *Nemacheilus* sp., has been made by Dhar & Majidah (1987).

The studies on the seasonal population of nematode parasites of fishes at two Sites of Wular Lake in Kashmir, *viz.* Ningli and Watlab, have been made by Majidah & Khan (1998), showing nematode parasite infection due to the Species like *C. cotti, R. himalayai* and *R. kashmirensis* in 6 spp. of Wular Lake fishes: *S. niger, S. curvifrons (=S.hugelii), S. esocinus. S. richardsonii (= Oreinus plagiostomus), Cyprinus carpio specularis* and T*riplophysa kashmirensis (= Nemacheilus kashmirensis).*

Recently, Chishti *et al.* (2002) conducted detailed investigations on the comparative fish parasite diversity, including nematode diversity in two fresh water lakes of Kashmir as urban eutrophic Dal Lake and rural mesotrophic Manasbal Lake, in relation to ecological factors in different seasons. The fish hosts studied during the course of three years were *S. curvifrons* and *Schizothorax richardsonii (= Oreinus sinuatus).* The nematode Species recorded were *Camallanus oxycephalus* and *R. guptai.*

The above authors have found that the nematode infection in *Schizothorax* showed existence in both the waterbodies, showing peak availability towards winter and spring season. The infection due to nematode parasite, *C. oxycephalus* occurred during July and August, with resulting peak in population density during late summer and autumn. Further, the parasite richness revealed that fishes of Dal Lake were heavily parasitized due to perhaps eutrophication.

4.1.2.3. Amphibians and reptilian nematodes

In Kashmir region, amphibian, *viz., Bufo viridis* (toad) has been found to be parasitized by four species of nematodes in intestine part of alimentary canal. These species are: *Cosmocerca crenshacwi, C. commutata (= C. kashmirensis)* (cosmocercidae). *Oswaldcruzia kashmirensis* (Molineidae); *Oxysomatium srinagarensis* (Oxyuridae). Besides these species, *Rhabdias bufonis* (Rhabdiasidae), is reported to be as a common lung parasite of toad in Kashmir Valley.

Two snake nematode species: *Kalicephalus indicus* and *K. kashmirensis,* have been found in the stomach and oesophagus of *Topidontus platyceps* and *Ptyas mucosus* respectively, in Kashmir Valley. These nematode parasites are belonging to the order Strongylida (the bursate nematodes) and family Diaphanocephalidae. In addition to these, two more nematode species of reptile, *viz. Thelandros baylisi* and *T. jummensis,* belonging to the order Oxyurida family Pharyngodonidae. *T. baylisi* has been reported to parasitize lizards: *Agama tuberculata* and *Calotes versicolor* in Jammu and Kashmir, whereas, *T. jammuensis* is found to be parasitic on lizard (*Hemidactylus flaviviridis)* in Jammu region.

4.1.2.4. Birds (wild and poultry) nematodes

Eight Species of nematodes, *viz,. Acuaria skrjabini, (*Acuariidae), *Contracaecum milvi* (Anisakidae) *Diplotriaena buckelyi* (Diplotriaenidae), *Rusguniella turdi* (Acuariidae), *Syngamus trachea* (Syngamidae),

Amidostomum fulicae, Epomidiostomum uncinatum, E. anatinum (Trichostrongylidae), have been reported from Kashmir, infecting wild birds such as *Passer domestica, Milvus migrans lineatus, Acridotheres tristis, Turdus unicolor, Corvus splendens, Fulica atra atra* and *Anasplatyrhynchos* respectively. The site of infection due to these nematode species, has been observed to be as gizzards caecum, body cavity, stomach and trachea, etc. (Table: 4.3 p. 66).

There are about nine Species of nematode parasites, known to infect poultry birds like domestic fowl (indigenous), *Gallus domesticus* and *Anser domesticus* in Kashmir Valley. The locations of infection by these nematode species have been found as: small intestine, crop, oesophagus, gizzard, caecum and trachea. The nematodes of poultry importance of Kashmir include *Ascardia galli* (Ascarididae), *Capillaria annulata, C. caudinflata* (Trichuridae), *Echinuria uncinata* (Acuariidae), *Heterakis gallinarum, H. indica* (Heterakidae), *Sterptocara indica* (Acuariidae) and *Syngamus trachea* (Syngamidae) and *Amidostomum anseris* (Trichostrongylidae), see table 4.3, page 66.

Detailed investigations on prevalence and seasonal variations of nematode parasites of domestic fowl (*Gallus*) (indigenous), have been done by Fotedar & Khateeb (1987) and Pandit *et al.* (1991). The former authors studies about 140 local fowls, collected from different localities and areas of Kashmir Valley and thoroughly examined their gizzards, small intestine, colon and caecum, and reported the existence of two common intestinal (small) and caecal nematode parasites as *Ascaridia galli* and *Heterakis gallinarum (= H. gallinae)* respectively, of poultry importance. The above authors have shown the highest incidence of infection in the month of September and lowest in the month of December and January. It was also reported by them that the incidence and mean worm burden indicated a decrease with decreasing temperature and rainfall.

Pandit *et al.* (1991) studied the prevalence of helminth parasites, including nematodes of indigenous fowls of Kashmir Valley after examining about 80 birds and it was found that about seven Species of nematode parasites cause much damage to poultry in this region, and they also become susceptible to other diseases. The seven Species of nematodes found by them, causing infections in different body sites/ systems of birds (domestic fowl), with their rate of infection were: *A. galli* (small intestine), 28%; *H. gallinarum* (caecae), 22%; *Syngamus trachea* (trachea), 9%;*Heterakis indica* (caecae); 8%; *Capillaria annulata* (crop and oesophagus), 4%; *C. caudinflata* (small intestine), 2% and *Echinuria uncinate* (gizzard), 1%. The most infections reported was single, however, mixed infections with two or three species was not uncommon. The occurrence of these parasites was described by these authors as greatly influenced by climatic and ecological factors of the region.

4.1.3. Rat, shrew, bat and cat nematodes

In Kashmir Valley, two Species of parasitic nematode, *viz, Heterakis spumosa (= Ganguleterakis spumosa)* and *Trichocephalus muris*, infecting intestine and stomach respectively, of common rat (*Rattus rattus),* have so far been reported. The domestic cat (*Felis domestica)* has been found to be infected in this Valley by a cosmopolitan nematode parasite of felines, *viz. Toxocara cati (= T.mystax)* (see table 4.3. page 67*)*

The capillariinid nematodes: *Capillarina minuta and Capillaria* sp., have been found to infect oesophagus and stomach of shrew (*Crocidura* sp.) of Kashmir region. In addition to these, *Capillaria kashmirensis* is reported to infect the stomach of bat, *Myotis muricola* (Vespertilionidae) from Anantnag district of Kashmir (see table 4.3, page 67).

4.1.4 Ovine, bovid, yak and hangul nematodes

4.1.4.1 Ovine nematodes

A total of about 22 species of nematodes, which commonly infect gastro-intestinal tract, abomasum and lungs of sheep, goat and lambs, have so far been recorded from various regions and localities of Jammu, Kashmir and Ladakh region. These nematode species are belonging to 10 genera, distributed over 8 different families, under five distinct orders: (Rhabditida, Strongylida Oxyurida, Ascarida (Class Secernetea) and Enoplida (Class Adenophora).

The 22 nematode perasite Species, covering many families, infect ovines of this region are: Ancylostomatids–*Bunostomum triganocephalum, Monodontus alatus*; Ascarid–*Ascaris lumbricoides*; Chabertiids–*Chabertia himalayana, Chabertia ovina, Oesophagostomum columbianum*; oxyurid–*Skrjabinema ovis*; Protostrongylids–*Muellerius* sp, *Protostrongylus rufescens, Varestrongylus pneumonicus Varestrongylus* sp.; Strongyloidid–*Strongyloides* sp; Trichostrongylids–*Dictyocaulus filaria, Haemonchus contortus, H. C. kashmirensis, Marshallagia marshalli, Nematodirus fotedari, N. spathiger, Trichostrongylus axei, T. colubriformis, T. trichostrongylus* sp. and Trichurid–*Trichuris ovis.*

The earliest detailed studies on gastro–intestinal nematode parasites in sheep of Kashmir, have been done by Dhar *et al.* (1982). They have revealed a total of 8 Species of nematodes belonging to genera, *viz. Bunostonum, Chabertia, Haemonchus, Nematodirus, Skrjabinema, Trichostrongylus* and *Trichuris,* infecting abomasum and intestinal tract of both sexes of sheep. The infection with *Trichostrongylus axei* and *T. colubriformis* was 57.7 and 40.2% respectively. Of the total nematode species recovered, the most common infection by *Trichostrongylus* spp. together accounted for 97.9% of the total worms recovered at necropsy. It was further, shown by the above authors that lambs in Kashmir region start building up heavy infection from May onwards reaching maximum levels during September.

Studies on population kinetics, development of juvenile stages and histopathology of bursate nematodes of Kashmir, *viz. Oesophagostomum columbianum* and *Chabertia himalayanum,* common intestinal parasites of sheep and lambs in Kashmir, have been accomplished by Bali & Fotedar (1974), Sharma & Dhar (1978, 79) and Fotedar & Wali (1984). *O. columbianum,* commonly known as nodule worm, reported to cause disease known as oesophagostomomiasis. The larvae of nematode parasite of *O. columbianum* cause lesions in the intestinal mucosa and nodules like dense capsules are formed out of these lesions. This disease can lead to chronic diarrhoea, the animal become weak and may even die. The adult worms of *Chabertia* attack intestine of sheep.

Baqui & Shah (2001), reported the incidence and intensity of nematode Species such as: *Haemonchus controtus, Marshallegia marshalli, Dictyocaulus filariae, Trichuris ovis* and *Bunostomum triganocephalum,* in ovines of Kashmir, during Januay to December 2002. They also studied the alteration of haematological parameter in ovine nematodiasis, such as Hb%. EST, PCU. MCH, Eosinophil and RBC counts, in relation to worm burden and faecal egg counts. The nematode infection exerted a great impact on the haematological picture of ovine nematodiasis.

- **Ascariasis (sheep):** The high incidence of ascarid infection in sheep in Handwara district of Kashmir, due to nematode parasite, *Ascaris lumbricoides* was noticed by Dhar *et al.* (1983). This finding was further supported from the information collected by the authors from medical hospital, which confirmed high incidence of human ascariasis at Handwara region and in the adjoining area. Sheep actually acted as an important agent for transmitting the human Ascarid infection in the area.

- **Lungworms of Ovines:** The various species of lungworms, infecting domestic ovines, occurring in different localities and regions of Kashmir and Ladakh are: *Dictyocaulus filaria* (Trichostronglyloidea: Trichostromgylidae); *Muellerius* sp., *Protostrongylus rufuscens*, *Varestrongylus pneumonicus* and *Varestrongylus* sp., belonging to super family Metastrongyloidea and family Protostrongylidae.

Dictyocaulus harbours the bronchial tubes of sheep, are thread like worm and having length of several inches. These worms cause coughing and bronchitis in hosts, and may block off so much of the lungs as to be fatal. *Muellerius* is a cosmopolitan nematode parasite of sheep and goat, and attacks the parenchyma of lung of these domestic animals. *Protostrongylus rufuscens*, usually occurs in the trachea, bronchi and bronchioles of sheep and goat. This lungworm is also called as red-lungworm, because of its colour due to ingested blood. *P. rufuscens* is pathogenic and its infection can lead to fatal pneumonia. The Species of the genus *Varestrongylus*, are parasites of lungs of domesticated sheep and goats, infecting bronchi and bronchioles of these animals of veterinary importance.

In various districts and tehsils of Kashmir and Ladakh, detailed investigations on the prevalence, distribution, pathology and control of lungworm infections in sheep and goat, have been conducted by: Dhar *et al.* (1972, 79); Dhar & Sharma (1978, 78a, 79); Dhar (1980, 82,85); Bansal *et al.* (1982) & Sharma (1994).

In Kashmir Valley, Dhar *et al.* (1972), were the first authors to study in detail the prevalence of lungworm diseases in sheep. For this purpose, they selected the animal farms of Government as well as those belonging to private breeders. The areas of study included various localities of north-east parts of Kashmir, *viz.* Dachigam, Shuhama, Goabal (Kangan), Chervan and Gund. It was reported that lungworm, *Dictyocaulus filaria* infected 21.3% of lungs and *Varestrongylus* sp. 29.5% of the lungs examined. Mixed infections in 6.6% of the lungs, having *D. filaria* and *Varestrongylus* sp, was also encountered. Apart from this, infection with *Varestrongylus* was found to be more common as compared to *D. filaria*.

Further studies on the prevalence of infection due to *D. filaria* in Kashmir were conducted by Dhar & Sharma (1978, 78a). They reported the prevalence of infection to be more in goat than in sheep. Later, Dhar *et al.* (1979) showed the occurrence of another lungworm disease caused by nematode, *Muellerius* sp. in sheep of Kashmir Valley.

The epidemiological studies on lungworms in Kargil region of Ladakh, have been carried out by Dhar & Sharma (1979). Their study has revealed higher prevalence of lungworm infection in goats than in sheep. The ovines (domesticated) of the Pandras region, showed highest prevalence of infection than those of the Government Farm at Khumbathang. The common lungworm of sheep and goat in Kargil, reported by Sharma (1994), included *D. filaria, Protostrongylus rufescens, Varestrongylus pneumonicus* and occasionally *Muellerius* sp. The prevalence of lungworm disease in ovines due to *Dictyocaulus viviparous* nematode parasite, for the first time was reported from Lolab area of Kashmir valley by Sharma & Dhar (1987).

In a comprehensive review paper on bovine and ovine parasitic bronchitis, with particular reference to the research work done in India, including J & K State, Dhar (1985) has concluded that lungworm disease (Verminous pneumonia) is very economically important disease of cattle and sheep in this region. Further, according to this author, sheep and goat are known to be infected by more than one species of lungworm parasite, more sheep of private breeder as compared to Government Farms, have been found infected with lungworms and, autumn and winter months showed high incidence and intensity of lungworm infection in the animals, mostly young ones.

4.1.4.2 Bovid nematodes

In Jammu region, the incidence of nematode parasites, *viz. Parafilaria bovicola*, infecting bullock and *P. sahii*, parasite of buffalo, belonging to the family Filariidae, sub-family Filariinae, were reported by Khajuria (1966). These nematodes were responsible for causing haemorrhagic filariasis in bovids. The disease caused formation of sub-cutaneous nodules, mostly on the neck or back regions and would break open and bleed profusely for months together. The rate of infection was found high during summer and rainy season.

Bhadwal & Raina (1994), reported the existence of nematode parasite–*Thelazia* (Thelazioidae: Thelaziidae: Thelaziinae), causing disease in buffalo in J&K known as thelaziasis. In this disease, the parasite affect eye orbit under the lids, conjunctiva and nictitating membrane, and lachrymal glands and ducts of domestic mammal. The above authors mechanically removed 6 worms from left eye and 11 worms from right eye in 6 year buffalo. The animal was suffering from wound infection of eye.

The bovine parasitic brochitis disease in bovines of this region is caused by a single Species of lungworm parasite, *viz. Dictycaulus viviparous*. The parasite mostly affected the young animals and disease showed high prevalence and intensity during autumn and winter months (Dhar, 1985)

4.1.4.3 Yak *(Bos grunniens)* nematodes

Ranga Rao (1994), studied the nematode parasites of Indian Yak (*B. grunniens*) in Ladakh region of J&K State. They found 10 spp. of gastro-intestinal nematode parasites and 1 sp. from peritoneum of Yak. These species were: *Ascaris* sp. (Ascarididae); *Bunostomum* sp., *Chabertia* sp. (Ancylostomatidae); *Cooperia hrenktahnsa, Haemonchus controtus, Mecistocirrus digitatus, Nematodirus* sp. *Ostertagia* sp. (Trichostrongylidae); *Setaria cervi* (Onchocercidae); *Toxocara vitulorum* (= *Neoascaris vitulorum*) (Ascarididae) and *Trichostrongylus* sp. (Trichostrongylidae). Excepting *S. cervi*, found in peritoneum, all other nematode Species were found to be infecting the gastro-intestinal tract of *B. grunniens*. (Table 4.3, pages. 67-68).

4.1.4.4. Hangul *(Kashmir Stag)* nematodes

The unique and endangered species of mammalian wildlife, *viz. Cervus elaphus hanglu*, has been found to be infected by seven common nematode parasites of gastro-intestinal tract. Shahader *et al.* (1995), for the first time reported the prevalence these parasites in Hangul wildlife of Kashmir and have identified these parasitic nematodes upto only genus level. The nematode parasite genera included *Haemonchus, Marshallagia, Nematodirus, Oesophagostomum, Ostertagia, Strongyloids and Trichostrongylus.* These parasites are responsible for causing ill health in the wild animal besides, animal becomes susceptible to other diseases (Table 4.3, page 69).

4.1.5 Human nematodes

4.1.5.1 General nematode parasites

Arora *et al.* (1976), were the first workers to conduct detailed survey of intestinal parasites in rural communities in Jammu region. They found four important parasites, *viz. Ascaris lumbricoides*, Hookworm,

Entrobius vermicularis and *Trichuris trichura*, in rural population in Lambri village of Rajouri district of Jammu. *A. lumbricoides* and hookworms were found in 6.0 and 4.6 percent respectively of the population. It was also observed that significantly higher prevalence of hookworm infection existed in females of higher age group (15 years and above) of the population .The first report of human toxocariasis due to nematode, *viz. Toxocara canis* in Kashmir, has been given by Ahmad *et al.* (2002). *T. canis* is a parasite of canines worldwide.The infection in human is caused due to accidental igestion of embryonated eggs containing second stage larvae.The migrating larvae result in serious clinical manifestations,leading to high morbidity and mortality in human. These authors analysed the antibody response to purified excretory-secretory antigen of *Toxocara canis* in patients harbouring *A. lumbricoides* and relevant control, to define, the extent of human toxocariasis in Kashmir region,being the highly endemic area of human ascariasis in the country.This study has shown,38(82.60%),out of 46 ascariasis patients,gave positive response to ES antigen.

4.1.5.2 Ascariasis

Wani (1970) for the first time conducted a thorough research study on incidence and symptomology of ascariasis due to nematode parasite, *Ascaris lumbricoides* in human population of Kashmir. He found 58% of positive cases of ascariasis in patients who attended S. M. H. S. Hospital (Government College), Srinagar. Further, the high percentage about 80.7% was observed in age group 6-10 years and lesser incidence in higher age group.

In the later years, Ahmad (1978) made investigations on the incidence of *A. lumbricoides* in the rural population of Kashmir. He revealed that of the 3 intestinal parasites (*A. lumbricoides*, *Gardia lambia* and *Taenia saginta*), *A. lumbricoides* was very much predominant and accounted for 61.8% of total cases examined. The incidence of ascariasis disease among males and females was notices 64% and 59% respectively. The declining trend was further seen with rise in adult age after 15 years. The higher incidence of 89.2% and 73.7% of parasite infestation was found in Kashmir rural population, covering school going age groups and pre-school age groups respectively.

The higher incidence of nematode parasite infection among the human population of Kashmir region, is believed to be because of faeco-oral contact of poor personal hygiene and also due to extensive contamination of soil and water in different localities and areas of rural regions of Kashmir. Yaseen (1979) studied the intestinal functions of ascariasis in Kashmir valley and has concluded that *Ascaris* infestation does not lead to functional or structural impairment of small intestine, resulting in mal-absorption syndrome.

Intestinal ascariasis is highly endemic in all parts of Kashmir besides, surgical, hepato-biliary, gall-bladder and pancreatic ascariasis, due to *A. lumbricoides*, have also been widely reported from different regions of Kashmir(Khuroo,1996). In Kashmir, several studies have been done on: biliary ascariasis by Khuroo & Zargar (1985), Hangloo *et al.* (1989, 92) and Wani & Chrungoo (1992); Hepato-biliary ascariasis by Khuroo (1999); pancreatic ascariasis by Khuroo *et al.* (1992 a) and Khuroo (1999); gall-bladder ascariasis by Khuroo *et al.* (1992); surgical ascariasis by Mir *et al.* (2002).

The hepato-biliary and pancreatic ascariasis (HPA), is a disease of great public importance in endemic areas of Kashmir. In HPA, the ascarides make repeated visits into and out of the bile and pancreatic duct. During its repeated but limited exploration of papillary orifice and advance into ducts, the person comes down with severe hepato-biliary or pancreatic disease. The dead ascarides in the duct lead to the formation of brown pigment 'stones', leading to the syndrome of Recurrent Pyogenic Cholangitis, a disease more often in Oriental region known as Cholangiohepatis (Khuroo, 1999).

The Species richness, systematic position and host diversity of nematode parasites, infecting insects (Invertebrata) and fishes, amphibians, wildbirds, poultry, small mammals, domestic animals,

TABLE 4.1. Species richness and host diversity of nematodes parasites, infecting invertebrate (Insects) and vertebrates

S. No.	Nematode genus	No. of Species	Number of species of nematodes in various host group									
			IN	FI	AM	RE	WB	PO	SM	DA	WL	HU
1.	*Amidostomum*	2					1	1				
2.	*Ascardia*	1						1				
3.	*Ascaris*	2								2		1
4.	*Acuaria*	1					1					
5.	*Bunostomum*	2								2		
6.	*Camallanus*	3		3								1
7.	*Capillaria*	5						2	3			
8.	*Chabertia*	3								3		
9.	*Cooperia*	1								1		
10.	*Controcaecum*	1					1					
11.	*Cosmocerca*	2			2							
12.	*Dictyocaulus*	2								2		
13.	*Diplotriaena*	1					1					
14.	*Echinuria*	1						1				
15.	*Entrobius*	1										1
16.	*Epomidiostomum*	2					2					
17.	*Galebilla*	2	2									
18.	*Haemonchus*	2								2	1	

Table contd....

S. No.	Nematode genus	No. of Species	Number of species of Nematodes in various host group									
			IN	FI	AM	RE	WB	PO	SM	DA	WL	HU
19.	*Heterakis*	2						2	1			
20.	*Hookworm*	1										1
21.	*Kalicephalus*	2				2						
22.	*Marshallagia*	2								2	1	
23.	*Mecistocirrus*	1								1		
24.	*Monodontus*	1								1		
25.	*Muellerius*	1								1		
26.	*Nematodirus*	4								4	1	
27.	*Oesophagortomum*	2								2	1	
28.	*Ostertagia*	2								2	1	
29.	*Oswaldcruzia*	1			1							
30.	*Oxysomatium*	1			1							
31.	*Parafilaria*	2								2		
32.	*Procamallanus*	1		1								
33.	*Prostrongylus*	1								1		
34.	*Pseudonymus*	1	1									
35.	*Rhabdias*	1			1							
36.	*Rhabdochona*	4		4								
37.	*Rusguniella*	1					1					

Table contd...

S. No.	Nematode Genus	No. of Species	Number of species of nematodes in various host group									
			IN	FI	AM	RE	WB	PO	SM	DA	WL	HU
38.	*Setaria*	1								1		
39.	*Skrajabinema*	1								1		
40.	*Spinitectus*	1		1								
41.	*Streptocara*	1						1				
42.	*Strongyloides*	2								2	1	
43	*Syngamus*	1					1	1				
44.	*Thelandros*	2				2						
45.	*Thelazia*	1								1		
46.	*Toxocara*	1							1	1		1
47.	*Trichocephalus*	1							1			
48.	*Trichostongylus*	4								4	1	
49.	*Trichuris*	2								1		1
50.	*Varestrongylus*	2								2		
	Total Number	**85**	03	09	05	04	08	09	06	41	07	04

Abbreviations used; IN = Insects; FI = Fishes; AM = Amphibians; RE = Reptiles; WB = Wildbirds; PO = Poultry; SM = Small mammals; DA = Domestic animals; WL = Wildlife (Hangul) and HU = Human.

TABLE 4.2. Systematic position and host diversity of nematode parasites of vertebrates (fishes, amphibians, reptiles, birds and mammals) of J & K State.

(I) Phylum: Nematoda; Class Secernetea

Order/sub-order (a)	Super family (b)	Family (c)	Sub-family (d)	Nematode genus (e)	Host (f)
Rhalditida	Rhalditiodea	Rhaldiasidae	—	*Rhabdias*	AM
-do-	-do-	Strongyloididae	—	*Strongyloides*	OV, HA
Strongylida	Diaphanocephaloidea	Diaphanocep-	—	*Kalicephalus*	SN
	halidae				
-do-	Anclostomatoidea	Ancylostomatidae	Ancylostomalinae	*Ancylostoma*	HU
-do-	-do-	-do-	Bunostominae	*Bunostomum*	SP, VA
-do-	-do-	-do-	-do-	*Monodontus*	SP
-do	Strongyloidea	Chabertiidae	Chalertiinae	*Chabertia*	SP, VA
-do-	-do-	-do-	Oesophago-stominae	*Oesophagostomum*	OV, HA
-do-	-do-	-do-	Ostertaginae	*Ostertagia*	YA, HA
-do-	-do-	-do-	-do-	*Marshallagia*	SP, HA
-do-	-do-	-do-	Trichostrongylinae	*Trichostrongylus*	SP, HA, OV
-do-	-do-	-do-	Haemonchinae	*Haemonchus*	SP, YA, HA
-do-	-do-	-do-	Mecistocirrinae	*Mecistocirrus*	YA
-do-	-do-	-do-	Dictyocaulinae	*Dictyocaulus*	BO, OV
-do-	-do	-do-	Amidostomanie	*Amidostonum* *Epomidiostomum* *Oswaldcruzia*	WB, PO WB, AM
-do-	-do-	-do-	Nematodirinae	*Nematodirus*	SP, YA, HA

Table contd...

(a)	(b)	(c)	(d)	(e)	(f)
-do-	Metastrongyloidea	Protostrongylidae	—	*Muellerius*	OV
-do-	-do-	-do-	—	*Protostrongylus*	GO
-do-	-do-	-do-	—	*Varestrongylus*	OV
-do-	-do-	Syngamidae	Syngaminae	*Syngamus*	WB, PO
-do-	Trichostrongyloidea	Trichostrongylidea	Cooperiinae	*Cooperia*	YA
Oxyurida	Oxyuridea	Oxyuridae	—	*Enterobius*	HU
-do-	-do-	-do-	—	*Skrjabinema*	SP
-do-	-do-	-do-	Oxysomatiinae	*Oxysomatium*	AM
-do-	-do-	Pharyngodonidae	—	*Thelandros*	LI
Ascaridida	Cosmocercoidea	Cosmocercidae	—	*Cosmocerca*	AM
-do-	Heterakoidea	Heterakidae	Heterakinae	*Heterakis*	PO, RA
-do-	-do-	Ascaridiidae	—	*Ascaridia*	PO
-do-	Ascaridoidea	Anisakidae	Anisakinae	*Contracaecum*	WB
-do-	-do-	Ascarididae	Ascaridinae	*Ascaris*	SP, HU
-do-	-do-	-do-	Toxocarinae	*Toxocara*	CA,HU
Spirurida/ Camallanina	Camallanoidea	Camallanidae	—	*Camallanus*	FI
-do-	-do-	-do-	—	*Procamallanus*	FI
Spirurida / Spirurina	Thelazoidea	Thelaziidae	Thelaziinae	*Thelazia*	BO
-do- / -do-	-do-	Rhabdochonidae	—	*Rhabdochona*	FI
-do- / -do-	Habronematoidea	Cystidicolidae	—	*Spinitectus*	FI
-do- / -do-	Acuarioidea	Acuariidae	—	*Acuaria*	WB
-do- / -do-	-do-	-do-	—	*Rusguniella*	WB
-do- / -do-	-do-	-do-	Seuratiinae	*Streptocara*	PO
-do- / -do-	-do-	-do-	Acuariinae	*Echinuria*	PO

Table contd...

Spirurida/Spirurina -do- / -do-	Filarioidea -do-	Filariidae Onchocercidae	Filariinae Setariinae	*Parafilaria* *Setaria*	BO YA
-do- / -do-	Diplotriaenoidea	Diplotriaenidae	Diplotriaeninae	*Diplotriaena*	WB
(II) Phylum Nematoda; Class Adenophora					
Enoplida/Trichinella	Trichinelloidea	Trichuridae	Trichurinae	*Trichuris*	SP, HU
-do- / -do-	-do-	-do-	-do-	*Trichocephalus*	RA
-do- / -do-	-do-	-do-	Capillariinae	*Capillaria*	BA, SH

Abbreviations used: AM = Amphibian (toad); BA = Bat; BO = Bovids; CA = Cat; FI = Fishes; GO = Goat; HA = Hangul (Wildlife); HU = Human;

LI = Lizards; OV = Ovines (Sheep, goat and lambs); PO = Poultry birds; RA = Rat; SH = Shrew; SP = Sheep; SN = Snakes; WB = Wild birds and YA = Yak (Indian).

For general classification and nomenclature of nematodes, Chitwood (1950), Yamaguti (1961 & 85), D' Hondt (1997) and Anderson (2000), have been followed.

TABLE 4.3 Animal nematodes of invertebrate (insects) and vertebrates: An up-to-date catalogue

Host group/Nematode species (a)	**Host species (b)**	**Location of infection (c)**	**Locality (d)**
(I) Invertebrate (Insects)			
1. *Galebilla galebiella*	Aquatic beetle	Intestine	Kashmir (Hill stream)
2. *Galebilla islamabadi*	-do-	-do-	-do-
3. *Pseudonymus multiannulata*	*Hydrophilus* sp.	-do-	Kashmir
(II) Vertebrates **FISHES** 1. *Camallanus cotti* (=*Camallanus fotedari*) *Schizothorax curvifrons*, *S. esocinus*,	*Cyprinus carpio*, *Nemachilus kashmirensis*, *Oreinus plagiostomus*; *Schizothorax curvifrons S. esocinus*	Intestine	Kashmir Valley Lakes (Wular, etc.)

Table contd...

(a)	(b)	(c)	(d)
2. *Camallanus oxycephalus*	*S.hugelii, S. niger*	Intestine	Kashmir (Dal,-Manasbal)
3. *Camallanus solmonae*	*Schizothorax* spp.	Intestine	Kashmir.
4. *Procamallanus (Monospiculus)-kashmirensis*	*Salmo* sp.*Wallago attu*	Intestine	Jammu
5. *Rhabdochona guptai*	*Cyprinus C. spicularis Orenius sinuatus* and *S.curvifronts*	Intestine	Hokarsar, Dal lake, Manasbal
6. *Rhabdochona (Filochona) kashmirensis*	*C.c. spicularis,*Nemachilus kashmirensis, *Orienus plagiostomus S. curvifrons and S. hugelii*	Intestine	Kashmir Valley
7. *Rhabdochona (R.) himalayai*	*C.c spicularis,* *Nemachilus* sp., *O. plagiostomus, S. esocinus, S. curvifrons and *S.hugelii*	Intestine	Lakes (Wular, etc.) Kashmir Valley
8. *Rhabdochona* sp.	*Salmo trutto fario*	Intestine	Kashmir
9. *Spinitectus* sp.	**Oreinus richardsonii*	Intestine	Jammu
(B) AMPHIBIAN			
1. *Cosmocerca crenshawi*	*Bufo viridis*	Intestine	Kashmir
2. *Cosmocerca commutata* (=*Cosmocerca kashmirensis*)	-do- -do-		-do- -do-
3. *Oswaldcruzia kashmirensis*	-do-	-do-	-do-
4. *Oxysomatium srinagarensis*	-do-	Rectum	-do-
5. *Rhabdias bufonis*	-do-	Lungs	-do-
(C) REPTILES			
1.*Kalicephalus indicus*	Snake (*Topidontus platyceps*)	Stomach	Pahalgam
2. *Kalicephalus kashmirensis*	Snake (*Ptyas mucosus*)	Stomach & Oesophagus	Srinagar

Table contd...

(a)	(b)	(c)	(d)
3. *Thelandros baylisi*	*Calotes versicolor, Agama-tuberculata*	Stomach	Jammu & Kashmir
4. *Thelandros jummensis*	*Hemidactylus flaviviridis*	Not known	
(D) BIRDS			
1. *Acuaria skrjabini*	*Passer domestica*	Gizzard	Srinagar
2. *Amidostomum fulicae*	*Fulica atra atra*	-do-	Dal Lake
3. *Amidostomum anseris*	*Anser domesticus*	Intenstine	Gund
4. *Ascardia galli*	Domestic fowl (indeginous)	-do-	Srinagar
5. *Capillaria annulata*	-do- -do-	Crop	Kashmir
6. *Caprillaria caudinflata*	-do- -do-	Oesophagus	Kashmir
7. *Contracaecum milvi*	*Milvus migrans lineatus*	Stomach	
8. *Diplotriaena buckelyi*	*Acridotheres tristis*	Body cavity	Jammu
9. *Echinuria uncinata*	Domestic fowl (indeginous)	Gizzard	Kashmir
10. *Epomidiostomum anatinum*	*Anas platyrhynchos-platyrhynchos*	-do-	Hokersar
11. *Epomidiostomum uncinatum*	-do- -do-	Gizzard	-do-
12. *Heterakis gallinarum*	-do- -do-	Caecum	Caecum Kashmir
13. *Heterakis indica*	Domestic fowl (indeginous)	-do-	-do- -do-
14. Rusguniella (Rusguniodes) turdi	*Turdus unicolor*	Intestine	Kashmir
15. Streptocara indica	*Gallus domesticus*	Not known	-do-
16. Syngamus trachea	*Corvus splendens* and Domestic fowl (indigenous)	Tracheae	-do-

Table contd...

(a)	(b)	(c)	(d)
(E) **MAMMALS** (Rat, Shrew, Bat & Cat)			
1. *Capillaria kashmirensis*	*Myotis muricola* (Bat)	Stomach	Anantnag
2. *Capillaria minuta*	*Crocidura* sp. (Shrew) stomach	Oesphagus	Kashmir
3. *Capillaria sp.*	*-do- -do-*	-do-	-do-
4. *Heterakis ? speumosa (=Ganguleterakis spumosa)*	*Rattus rattus* (Rat)	Intestine	Kashmir
5. *Toxocara cati(=Toxocara mystax)*	*Felis domestica* (Cat)	Not known	Kashmir
6. *Trichocephalus muris*	*R. rattus.*	Stomach	-do-
(F) **DOMESTIC ANIMALS** (Sheep, Goat, Lamb, Yak, Buffalo, Bullock, Bovids, etc.)			
1. *Ascaris lumbricoides*	Sheep	Intestine	Kashmir
2. *Ascaris sp.*	Yak *(Bos grunniens)*	Gastro-intestine	Ladakh
3. *Bunostomum triganocephalum*	Sheep	Intestine	Kashmir
4. *Bunostomum* sp.	Indian Yak	Intestine	Ladakh
5. *Chabertia himalayana*	Sheep	Intestine	Kashmir
6. *Chabertia ovina*	Sheep	Gastro-	Handwara
7. *Chabertia* sp.	Indian Yak	Intestine	Ladakh
8. *Cooperia hrenktahnsa*	Indian Yak	-do-	-do-
9. *Dictycaulus filaria*	Goat, sheep	Lung	Kargil (Ladakh)
10. *Dictycaulus viviparous*	Domestic animal (bovids)	Lung	Lolab (Kashmir)
11. *Haemonchus contortus*		Intestine, stomach & abomasum	Handwara
12. *Haemonchus contortus var. kashmirensis*	Sheep, Yak	Gastro-intestine	Srinagar, Kashmir

Table contd...

(a)	(b)	(c)	(d)
13. *Marshallagia marshalli*	Sheep	Intestine	Kashmir
14. *Mecistocirrus digitatus*	Indian Yak	Gastro-intestine	Ladakh
15. *Monodontus (Pseudogaiheria) alatus*	Sheep	Not known	Kashmir
16. *Muellerius* sp.	Sheep, Goat	Lung	Kargil, Ladakh
17. *Nematodirus fotedari*	Sheep	Intenstine	Kashmir
18. *Nematodirus spathiger*	Sheep	Gastro-intestine	Handwara
19. *Nematodirus* sp.	Indian Yak	Gastro-intestine	Ladakh
20. *Oesophagostomum columbianum*	Sheep,lamb.	Intestine	Kashmir
21. *Ostertagia* sp.	Indian Yak	Gastro-intestine	Ladakh
22. *Paraifilaria bovicola*	Bullock	Sub-cutaneous nodules in neck/back region	Jammu
23. *Parafilaria sahaii*	Buffalo	-do-	-do-
24. *Prostostrongylus rufescens*	Goat	Lung	Ladakh (Kargil)
25. *Setaria cervi*	Indian Yak	Peritoneum	Ladakh
26. *Skrjabinema ovis*	Sheep	Gastro-intestine	Handwara
27. *Strongyloides* sp.	Ovines	-do-	Kashmir
28. *Toxocara vitulorum* (=*Neoascaris vitulorum*)	Indian Yak	Gastro-intestine	Ladakh
29. *Thelazia* sp.	Buffalo	Eyes	Jammu and Kashmir
30. *Trichostongylus axei*	Sheep	Abomasum & intestinal tract	Handwara
31. *Trichostongylus colubriformis*	Sheep	-do-	-do-

Table contd...

(a)	(b)	(c)	(d)
32. *Trichostongylus* spp.	Indian Yak, Ovines	Gastro-intestine	Ladakh
33. *Trichuris ovis*	Sheep	Abomasum & intestinal tract	Handwara
34. *Varestrongylus pneumonicus*	Sheep, Goat	Lung (bronchi/bronchioles)	Kashmir, Ladakh
35. *Varestrongylus* sp.	Sheep, Goat	Lung (-do-)	J & K State
(G) WILDLIFE *(Cervus elaphus hanglu)*	*C. elaphus hanglu* (Hangul)	*Gastro-intestine*	Kashmir
1. *Haemonchus* sp.	-do-	-do-	-do-
2. *Marshallagia* sp.	-do-	-do-	-do-
3. *Nematodirus* sp.	-do-	-do-	-do-
4. *Oesophagostomum* sp.	-do-	-do-	-do-
5. *Ostertagia* sp.	-do-	-do-	-do-
6. *Strongyloides*	-do-	-do-	-do-
7. *Trichostrongylus* sp.	-do-	-do-	-do-
(H) HUMAN			
1. *Ascaris lumbricoides*	Human	Gastro-intestinal tract	Jammu & Kashmir
2. *Ascaris lumbricoides hominis*	-do-	-do-	-do-
3. *Entrobius vermicularis*	-do-	Intestine	-do-
4. *Hookworm*	-do-	-do-	Rajouri (Jammu)
5. *Toxocara canis*	-do-	-do-	Kashmir
6. *Trichuris trichiura*	-do-	-do-	Jammu.

wildlife and human (Vertebrata), are encapsulated in the Tables: 4.1, pp. 59-61; 4.2, pp. 62-64.

Source (Table 4.3)

(i) Insect nematodes: Basir (1941); Fotedar (1964); Gupta (1967).

(ii) Fish Nematodes: Chakravorti (1942); Chishti & Bakshi (1990); Chishti *et al.* (2002); Dhar & Anwari (1982); Dhar & Fotedar (1979); Dhar & Majidah (1987); Fotedar & Dhar (1970, 75); Gupta (1967); Majidah & Khan (1996); Khan & Majidah (1974); Raina & Dhar (1972); Thapar (1950).

(iii) Amphbian Nematodes: Fotedar (1960, 65, 71, 73 and 80).

(iv) Reptile (snakes, lizard) nematodes: Fotedar *et al.* (1966, 68); Fotedar (1968).

(v) Poultry bird nematodes: Fotedar and Khateeb (1987); Pandit *et al.* (1991);Tanveer & Chishti (1999)

(vi) Wild bird Nematodes: Chishti (1978, 81) Chishti & Abdullah (1986); Fotedar & Chishti (1974, 80); Fotedar *et al.* (1972); Fotedar and Kaw (1965); Gupta (1967); Tanveer & Chishti (1999)

(vii) Rat and Shrew Nematodes: Raina (1970); Raina & Kaul (1974) and Raina & Raina (1985).

(viii) Cat Nematodes: Fotedar & Chishti (1974).

(ix) Bat nematode: Raina & Kaul (1982).

(x) Wildlife (mammal), Hangul (*Cervus elaphus hanglu*): Shahadar *et al.* (1995).

(xi) Indian Yak (*Bos grunniens*): Ranga Rao *et al.* (1994)

(xii) Domestic animals (mammals*)*:Bambroo (1971); Bali (1972); Bhadwal & Raina (1994); Baqui & Shah (2002); Dhar (1980, 82); Dhar & Sharma (1978, 78a, 79); Dhar *et al.* (1972, 79, 82, 82a, 83); Fotedar & Bambroo (1965, 68); Fotedar & Wali (1984); Gupta (1967); Khajuria (1966); Makhdomi *et al.* (1995); Sharma (1994); Sharma & Dhar (1978, 79, 87).

(xiii) Human nematodes*:* Arora *et al.* (1976) Hangloo *et al.* (1989); Khuroo(1999); Mir *et al.* (2002); Wani & Chrungoo (1992); Wattal *et al.* (1986); Ahmad *et al.* (2002).

4.2 PLANT-PARASITIC NEMATODES

4.2.1. General account, groups and damage caused by nematodes

Plant-parasitic nematodes belong to the group Nemathelminthes, class Nematodes. The nematode size ranges from 300-1000 m. They have eel-shaped unsegmented bodies, without legs and other appendages. The form of some Species, however, become swollen pear shaped or spheroid. Majority of plant-parasitic nematode feed on roots of different plant Species by feeding apparatus known as stylets, responsible for causing damage to higher plants, crops and other economically important plants, and also transmit virus diseases.

The plant-parasitic nematode diversity in subtropical and tropical countries of the world, have been studied by Luc (1990). In the recent years, the diversity of these economically important nematodes, affecting agricultural crops - food, vegetables, fruits, fibres and other economically important plants in Indian sub-continent, have been provided by Swarup & Das Gupta (1986), Chandniwala (1996) and Trivedi (1998 & 98a). The Jammu, Kashmir and Ladakh regions show great diversity of plant-parasitic nematodes. Many species have been reported to be as serious pests of temperate crops. However, detailed studies on a actual yield loss estimates have not been conducted thoroughly. In these regions, like other nematodes, plant- parasitic nematodes have successfully colonized in great variety of habitats.

The phytonematodes attack the roots of plants, through roots some of these parasites attack and damage aerial and foliage part of plant, such as stem, leaf, seed, causing gall - formation, lesions and rots, twisting or distortion of leaves and stem besides, stunting and abnormal development of floral parts. The nematodes infestation in plants and various crops leads to reduced plant vigour, yellowing of foliage, wilting, delayed flowering and low yield of poor quality.

On the basis of feeding habits, the various species of plant-parasitic nematodes, attacking valuable crops in J & K State, are grouped as ectoparasitc nematodes and endoparasitic nematodes. The ectoparasitic nematodes are external feeders and do not enter plant tissues, remain in soil and intermittently feed on roots, such as: Ring-nematodes (e.g., *Criconemella, Hemicriconemoides, Neolobocriconema*); Spiral-nematode (*Helicotylenchus, Rotylenchus*); Lance-nematodes (*Hoploliamus*); Needle-nematode(*Longiodorus*); pin-nematode (*Paratylenchus*); Stunt-nematode (*Tylenchorhynchus*); Dagger-nematode (*Xiphinema*) and Root-rot nematode (*Scutellonema*).

The endoparasitic nematode may be either semi–endoparasitic or completely endoparasitic. The semi-endoporasitic nematode species partially penetrate the root, head and neck, remain fixed in cortex and posterior region remain free in soil. These types of nematodes damage cortex, endosperm, pericycle and pholem parenchyma of its hosts.

The semi-endoparasitic nematodes include Reniform-nematodes (*Rotylenchulus reniform*). The endoparasitic nematodes enter host's roots during parasitization and these group of nematodes may be migratory endoparasitic nematodes and sedentary endoparasitic nematodes. The migratory endo-parasitic nematodes cover: Root–lesion nematodes, such as *Pratylenchus*; Burrowing nematode (*Radopholus*) Leaf or white-tip nematode (*Aphelenchoides*). Paddy – root nematode (*Hirschmanniella)* and Ear-cockle / seed or wheat-gall nematode (*Anguina tritici*). The sedentary endoparasitic nematodes of this region include Cyst–forming nematodes (*Heterodera* and *Globodera*) and Root – knot or gall – nematode (*Meloidogyne*).

Ectoparasitic Nematodes	Endoparasitic Nematodes
(I) Ring Nematodes	**A Semi-Endo Parasitic**
(i) *Criconemella*	(I) Reniform Nematode
(ii) *Hemicriconemodies*	(i) *Rotylenchulus*
(iii) *Neolobocriconema*	**B Migratory Endoparasitic**
(II) Spiral Nematodes	(I) Root – lesion Nematode
(i) *Helicontylenchus*	(i) *Pratylenchus*
(ii) *Rotylenchus*	(II) Ear-cockle/seed or Wheat-gall Nematode
(III) Lance Nematode	(i) *Anguina tritici*
i) *Hoplolaimus*	(III) Burrowing Nematode
(IV) Needle Nematode	(i) *Radopholus*

Table contd...

Ectoparasitic Nematodes	Endoparasitic Nematodes
(i) *Langidorus*	(IV) Leaf or White –tip Nematode
(V) Pin Nematode	(i) *Aphelenchoides*
(i) *Pratylenchus*	(V) Paddy-root Nematode
(VI) Stunt Nematode	(i) *Hischmanniella*
(i) *Tylenchorhynchus*	**C. Sedentary Endoparasitic**
(VII) Dagger Nematode	(I) Cyst-forming Nematodes
(i) *Xiphinema*	(i) *Heterodera*
(VIII) Root – rot Nematode	(ii) *Globedera*
(i) *Scutellonema*	(II) Root-Knot or gall Nematode
	(i) *Meloidogyne*

The major plant-parasitic nematode groups of agricultural crops of J&K, are given as under:

The number of serious plant-parasitic nematodes reported from J & K State regions, are known to cause considerable damage to various parts of number of valuable crops. These mainly include root damaging nematodes. and aerial or foliage part damaging nematodes. The nature and extent of damage rendered by these are given as under:.

The root-infesting nematodes like *Meloidogyne*, *Xiphinema* are responsible for producing characteristic galls on the roots of host plants. The nematodes, *viz. Pratylenchus, Radopholus* and *Neolobocriconema*, damage the cortex of roots by penetration and movement, and sometimes small lesions are also produced. *Rotylenchulus reniform* (Reniform-nematode) causes damage to cortex endosperm, pericycle and phloem parenchyma of roots. Sometimes nematode species, *viz. Meloidogyne hapla* and *Pratylenchus* spp. kill the distal part of root by girdling and remaining portion of root that is alive develops branch rootlets. This deformity is sometimes called as 'witches' broom 'or 'bearding'. *Xiphinema*, may at occasion causes root-surface necrosis in host-plant by killing the surface cells of large areas. *Meloidagyne* and *Xiphinema*, cause curly- tips in some cases, due to devitalize roots.

Root-rot nematodes–*Scutellonema* and *Radopholus* are responsible for extensive tissue damage, leading to rots by other organisms i.e. secondary invaders. The other root-attacking nematodes include *Heterodera* spp., *Hoplolaimus* spp., *Tylenchorhynchus* spp., *Rotylenchus* spp. and *Helicotylenchus* spp., widely occurring in many parts of J&K. States.

The above-ground infesting nematodes or aerial or foliage-parts damaging nematodes, are responsible for showing symptoms of different types. The stunt nematode (*Tylenchorhynchus*) causes stunted growth in plants and reduce the ability to withstand adverse condition. The nematode, *viz. Globodera* and *Aphenchoides*, cause nutritional deficiency in plants. The larval stages of gall- nematode (*Anguina*), live in growing points of wheat and develop distortion in stem and leaves. Also, *Anguina tritici* survives for long period in wheat cockles or galls.

4.2.2 Systematic survey and diversity of phytonematodes, affecting agricultural crops

The earlier authors who conducted surveys and studies on plant-parasitic nematodes of Kashmir, and also gave descriptions of new nematodes include Jairajpuri & Siddiqi (1963) Khan & Siddiqi (1963) Jairajpuri (1965) Jairajpuri and Baqri (1966), Raina (1966) and Jairajpuri & Loof (1967).

Mahajan (1971) for the first time gave a checklist of plant-parasitic nematodes of Jammu and Kashmir State, based on surveys made by him in different – major crop and fruit – growing areas and localities of the State. However, this list does not include data on specific genus and species of nematode, affecting the particular host-plant species.

In the recent past, several workers made thorough taxonomically studies, with providing descriptions of new species of phytonematodes of this region, affecting number of agricultural crops and economically important plants. In this connection, the valuable contribution have been made by : Mahajan (1972, 73) Fotedar & Mahajan (1971, 72,. 72 b, 73 & 74); Fotedar & Handoo (1974,77); Sheikh (1979) Gupta & Uma (1980 & 81), Fotedar & Kaul (1984,85); Kaul & Wali-ullah (1995) and Akhtar *et al.* (1996).

A list of Plant-parasitic nematodes, reported previously by various authors from Kashmir Valley, covering about 22 species of nematodes, affecting 12 spp. of crops and economically important plants, has been given by Fotedar & Kaul (1982). Further, their work also incorporated a checklist of 16 different genera of plant nematodes of Kashmir, with their distribution, population density, from 21 different localities, damaging agricultural crops (food, fruit, vegetable, oil seed) and, aromatic and medicinal plants. This list was mainly based on surveys conducted by them from 1979-81. In addition to this, these authors have also reported the wide distribution of nematodes, *viz.* Spiral-nematodes, belonging to the genus *Helicotylenchus,* Lesion-nematodes of the genus *Pratylenchus* and Stunt- nematode species , belonging to genus, *Tylenchorhynchus*.

Kaul (1998) studied the diversity and host-range of plant- parasitic nematodes, affecting vegetables, fruits and other crops. He indicated the existence of a total of 49 species of nematodes, belonging to 17 different genera in Jammu and Kashmir State. 30 nematode species have been shown to be attacking a wide variety of vegetable crops, 14 species affecting fruits crops, 10 spp. damaging food (cereal) crop, 2 spp. attacking pulse crop (*Phaseolus vulgaris*) and 1 sp. each affecting *Crocus sativus* (saffron), *Cynodon doctylon* and a grass species in J&K.

The data compiled during the course of present work, shows that a total of 93 spp. of phytonematodes, spread over 38 different genera belonging to order Tylenchida and Dorylaimida, are known to occur in Jammu, Kashmir and Ladakh region of Indian sub-continent (Table 4.4, pp. 78-86). These also include plant- parasitic nematodes, affecting various crops and other economically important plants of these regions.

The species diversity of phytoparasitic nematodes affecting different crops (cereals, fruits, vegetables, pulses) is as:cereals-paddy (9 spp.), maize (12 spp.), wheat (2 spp.); vegetables/pulses-brinjal (23 spp.), cabbage(5), cauliflower(9), chilly(10), french beans(5) soyabeans(1), kale(12), okra(10), onion(4), potato(19), pea(3), raddish(7), spinach(3), tomato(17), turnip(9) and winter squash(5); Fruit crops-apple(16), almond(6), apricot(1), cherry(12), peach(5), pear(3), pomgaranate(5), strawberry(4) and walnut(12).

The diversity, host range, distribution and population dynamics of various plant nematodes, affecting valuable agricultural crops (cereal, fruit and vegetable) of Jammu & Kashmir State, are briefly highlighted as under:-

4.2.3. Nematodes damaging agricultural crops (food, vegetables, fruits and mushroom)

(A) FOOD CROP

(i) Paddy*(Oryza sativa)* (Poaceae)*:* So far nine species of nematodes have been reported from Kashmir, damaging faddy, belonging to six orders. Mahajan (1971) was first worker to record nematode, *viz. Hirschmanniella mucronota* and *H. oryzae*, damaging paddy in Kashmir region. Wali-ullah (1991), during the course of his survey in various localities in districts (Srinagar, Badgam and Baramulla) of Kashmir, in connection with phytonematodes of *O. sativa var.* Ratna 39, encountered, the incidence of four species of nematode, belonging to geuus *Hirschmanniella*. It was found that *H. oryzae* was 71% in Badgam district followed by *H. gracilis*. In Srinagar district (Shalimar and Harvan locality), the author found the incidence of nematodes, *viz.. H. mucronota, H. gracilis* and *H. shemimi*, with population frequency 65%, 20% and 5% respectively. In Baramulla district, all the three species except *H. shamimi* were found by the author and these nematode species, showed varying frequencies and densities. Recently, Kaul & Wali-ullah (1995) reported the existence of *Tylenchorhynchus oryzae*, attacking rhizosphere of paddy roots in Kashmir. (See Table 4.4).

(ii) Maize *(Zea mays)* (Poaceae)*:* Twelve species of nematodes affecting *Z. mays* have till date been reported by various authors. The earliest recorded species of nematode of *Z. mays* from J & K were *Longidorus reneyii*, *Hoplolaimus* sp., *Helicotylenchus steineri*, *H. pteracercusoides*, studied by Raina (1996), Mahajan (1970), Fotedar & Mahajan (1974) and Fotedar & Kaul (1985) respectively. Later, Fotedar & Kaul (1982) recorded four more genera of nematodes, attacking maize plant in Kashmir. These included *Paratylenchus* sp. *Paurodontus* sp., *Pratylenchus* sp. and *Tylenchus* sp. The incidence of rest of the four species of nematodes, affecting maize crop in Kashmir Valley has been shown by : Fotedar & Kaul (1982) (*Longidorus* sp. and *Xiphinema*; Kaul (1987) (*Paratylenchus minimus*) and Kaul & Fotedar (1987) (*Helicotylenchus erythrinae*).See Table(4.4)

(iii) Wheat *(Triticum vulgare)* (Poaceae)*:* Fotedar & Mahajan (1973) were first to record the nematode, *viz. Basiria tritici* of wheat in Kashmir. Later, the occurrence of *Anguina tritici* (seed or wheat-gall nematode), causing 'Ear cockles' disease of wheat, was noticed by Sheikh (1979) in some wheat fields in the vicinity of Srinagar. The disease affected plant showed more or less stunted growth and leaves exhibited characteristics wrinkled, rolled or curled condition. In the recent years, the distribution of *A.tritici* in different wheat fields of various areas of Kashmir, have also been reported by Fotedar & Kaul (1982), Zaki & Mir (1994) and Zaki (1999), (See Table 4.4).

(B) VEGETABLE CROPS

(i) Brinjal *(Solanum melongena)* (Solanaceae)*:* As many as 23 species of nematodes, distributed over 11 different genera have so far been reported affecting Brinjal in J&K. These include 5 spp. of *Helicotylenchus*, 4 spp. of *Tylenchorhynchus*, 3 spp. of *Tylenchus*, 2 spp. each of *Meloidogyne* and *Pratylenchus,* and 1 sp.each of *Basiria, Hirschmanniella, Circonemella, Hoplolaimus*, *Nothotylenchus Paravulus* and *Xiphinema*.

The earliest records of nematodes, damaging Brinjal crop in Kashmir have been given by Fotedar & Kaul (1982).The important publications dealing with incidence and distribution of nematode-fauna, associated with Brinjal crop in Kashmir are given by: Fotedar & Kaul (1984, 85), Zaki *et al.* (1991) and Kaul (1998).(Table 4.4)

(ii) Cabbage *(Brassica oleracea var. capitata)* (Brassicaceae)*:* In all, five species of phytonemetods have so far been reported, affecting cabbage. These are: *viz. Helcotylcnchus indicus, H.silvactus, Hoplolaimus indicus, Tylenchorhynchus brassica, Tylenchus filiformis*. These species have been reported by Kaul (1982, 84 & 98), Siddiqi (1963) and Zaki *et al.* (1991) (Table 4.4).

(iii) **Cauliflower *(Brassice oleracea var. butrytis)* (Bassicaceae)*:*** A total of nine species of nematodes have been found to be affecting cauliflower vegetable. Each species belongs to genera, *viz. Basiri, Helicotylenchus, Hirsehmanniella, Nothotylenchus.* Besides, 2 species of *Pratylenchus* and three species of *Tylenchorhynchus,* have also been reported to damage the valuable vegetable crop by number of workers. (Table 4.4)

(iv) **Chilli *(Capsicum annuum* (Solaneceae)*:*** About ten different species of phytonematodes are damaging chilli crop in J and K. These cover: one species each of genus, *Helicotylenchus, Hirchmanniella, Paratylenchus*; 3 species of *Tylenchorhynchus*; 2 species each of *Meloidogyne* and *Tylenchus*. Zaki *et al.* (1991) have reported the incidence of *Meloidogyne hapla,* damaging severely chillies in the fields of Division of Olericulture (SKUAST), Shalimar, Kashmir (Table 4.4).

(v) **French Bean *(Phaseolus vlgaris)* (Fabaceae)*:*** Plant parasitic-nematodes, *viz. Circonemella* sp., *Helicotylenchus indicus, H. minzi, H. Helicotylenchus* sp., *Xiphinema* sp., are known to damage *P. vulgaris* in this region. These nematodes have been recorded by Fotedar & Kaul (1982) and Zaki *et al.* (1991). (See Table 4.4).

(vi) **Soya Bean *(Glycine max)* (Fabaceae)*:*** Nematode, viz. *Lordellonema bashirensis,* found in Safapore (Kashmir), inhabiting soil around the roots of *G.max.* (See Table 4.4).

(vii) **Kale *(Brassica oleracea var. acephala)* (Brassicaceae)*:*** A dozen of plant-parasitic nematodes have been found to damage the root of this local Cultivar. These include one species each of genera, *Basiria, Hirschmanniella, Hoplolaimus, Pratylenchus, Tylenchorhynchus* 3 spp. of *Helicotylenchus* and two spp. each of genera-*Nothotylenchus* and *Tylenchus*. These nematodes have been studied by Siddiqi (1962), Fotedar & Mahajan (1971, 72, 73 & 74), Fotedar & Kaul (1982), Zaki *et al.* (1991) and Walli-ullah (1992). (See Table 4.4).

(viii) **Okra *(Lady finger) (Ablemoschus esculentus)* (Malvaceae)*:*** As many as ten species of phytonematodes have so far been recorded from J&K, affecting okra vegetable in different areas and localities. These include one species each of genera, *viz. Basiria, Helicotylenchus, Hoplolaimus, Meloidogyne, Nothotylenchus, Pratylenchus* and *Rototylenculus,* besides three species of *Tylenchorhynchus* (See Table 4.4).

(ix) **Onion *(Allium cepa)* (Liliaceae)*:*** Only four species, *viz. B. kashmireusis H. dihystera, Helicotylenchus* sp. and *N. basiri,* are reported to affect. *A.cepa* in this region (Table : 4.4).

(x) ***Potato (Solanum tuberosum) (Solanaceae):*** About 19 species of nematodes have been found to be damaging potato in this region, distributed over 10 different genera. The various genera pertaining to these reported species are: *Aphelenchoides* (1sp.), *Basiria* (1sp.), *Globodera* (1sp.) *Helicotylenchus* (2 spp.) *Hoplolaimus* (1 sp.) *Meloidogyne* (2 spp.), *Pratylenchus* (5 spp.), *Tylenchrhynchus*(2 spp.), *Tylenchus* (3 spp.) and *Xiphinema* (1sp.) (Table 4.4)

High root population of number of nematodes attacking potato in wide field areas of Kashmir, have been studied by number of workers (Table: 4.4). It was Krishna- Prasad (1986) who for the first time recorded the existence of root – knot nematode (*Meloidogyne hepla*) from Kashmir. Of the various lesion nematodes, *Pratylenchus zeae* has been found to be a dominant species in the field by Wali-ullah & Bhat (1990).

A detailed survey conducted by Zaki *et al.* (1991) in potato crop fields of eight localities of Kashmir in the year 1988, revealed valuable information on population dynamics of nematode, parasitic on potato crop of this region. The maximum population of *Pratytenchus spp.* was noticed by them in early season and declining towards the harvest of potato i.e. August. According to these authors, *Tylenchorhynchus* spp. (Stunt-nematodes) showed high population during winter months, through October to April. In case of *Helicotylenchus* spp. (spiral-nematodes), maximum population was found by them during June and July months and minimum during December and January.

(xi) Pea *(Pisum sativum)* (Fabaceae)*:* In J&K, vegetable pea has been found to be affected by three important phytonematodes, *viz. Helicotylenchus dihystera, Pratylenchus brachyurus* and *Tylenchorhynchus mashhoodi*, reported by Kaul (1998), Wali-ullah & Bhat (1990) and Wali-ullah & Kaul (1997) respectively. (See Table 4.4).

(xii) Raddish *(Raphanus sativus)* (Cruciferae)*:* Raddish has been shown by many authors, affected by about 7 species of phytonematodes, belonging to seven different genera, *viz. Basiria Helilcotylenchus*, *Meloidogyne*, *Rotylenchus, Pratylenchus, Rototylenchulus* and *Tylenchorhgnchus* (See Table 4.4)

(xiii) Spinach *(Spinacea oleracea)* (Chenopodiaceae)*:* About three species of plant-parasitic—nematodes, *viz. Helicotylenchus, silvactus, Meloidogyne incognita* and *Tylenchus filiformis* have been reported to damage spinach in this region by Wali-ullah (1992) and Kaul (1998) (Table 4.4)

(xiv) Tomato *(Lycopersicum exulentum)* (Solanaceae)*:* About 17 species of plant-parasitic nematodes have so far been found affecting the tomato crop in this region by various workers. These include 5 spp. of *Helicotylenchus,* 4 spp. of *Tylenchorhynchus*, 3 spp. of *Tylenchus* and 1 sp. each of *Meloidogyne, Nothotylenchus, Pratylenchus, Scutellonema* and *Xiphinema.* (See Table 4.4).

(xv) Turnip *(Brassica campertris rapa)* (Brassicaceae)*:* Turnip crop in this region has been reported to be affected by 9 spp. of phytonematodes, including 1 spp. each *Helicotylenchus, Nothotylenchus Paratylenchus, Rototylencchus* and *Tylenchorhynchus* and 4 spp. of *Pratylenchus.* (See Table 4.4).

(xvi) Winter squash *(Cucurbita maxima)* (Cucurbitaceae)*:* The various phytonematode species damaging *C. maxima* in this State are: *Helicotylenchus* sp., *Meloidogyne incognita, Psilenchus, Rototylenchulus reniforms* and *Tylenchorhynchus* sp. (Table 4.4).

(C) FRUIT CROPS

(i) Apple *(Pyrus malus)* (Rosaceae)*:* About 16 spp. of phytoparasitic nematodes have been recorded by different authors, affecting the valuable apple crop in Kashmir, spread over 11 different genera, occurring in different localities and areas of this region. The highest number of species i.e. 3 belong to the genus *Helicotylenchus*, followed by genera *Enchodelus*, *Hoplolaimus* and *Tylenchus,* having 2 spp. each and rest of the seven genera have one 1 sp. each (see Table: 4.4)

(ii) Almond *(Prunus amygdalus)* (Rosaceae)*:* A total six species, belonging to 5 different genera of plant-parasitic nematodes, have been reported from J&K State, damaging almond fruit crop. These include two species of *Pratylenchus* and one each of rest of 4 genera. (Table: 4.4) Khan & Siddiqi (1963) were first to show the existence of nematode, *viz. Neolobocriconema laterale* (= (*Criconema laterale*) in Kashmir, affecting almond fruit crop.

(iii) Apricot *(Prunus armeniaca)* (Rosaceae)*:* An undescribed species of plant-parasitic nematode *Helicotylencus jammuensis,* has been reported of Mahajan (1971) from Jammu region, affecting the apricot (Table 4.4).

(iv) Cherry *(Prunus avium)* (Rosaceae)*:* About one dozen phytonematode species have been reported to damage cherry fruit tree in Kashmir by various authors, belonging to different genera. These include 3 spp. of genus *Helicotylenchus*, 2 spp. each of *Tylenchorhynchus* and *Xiphinema*, and 1 sp. each of *Aphelenchus*, *Rotylenchus*, *Criconemella*, *Meloidogyne and Pratylenehus*. Mahajan (1971) for the first time reported the existence of *Pratylenchus and Rotylenchus*, affecting cherry in Kashmir. In the recent years, during the course of faunistic survey of plant-parasitic nematodes of Cherry, more genera and species have been studied by Fotedar & Kaul (1982) and Wali-ullah & Kaul (1997). (See Table 4.4).

(v) Peach *(Prunus persica)* (Rosaceae)*:* 5 species of phytonematodes have so far been found to damage peach in J&K. These include *Helicotylenchus hazartbalensis*, *H .kashmirensis* sp., *Tylenchorhynchus pruni* and *Tylenchus filiformis*, reported by Fotedar & Handoo (1974), Gupta & Uma (1981) Fotedar & Kaul (1982) (Table 4.4).

(vi) Pear *(Prunus communis)* (Rosaceae)*:* According to Fotedar & Kaul (1982) three phytoparasitic nematodes, viz. *Helicotylenchus*, *Longidorus* and *Xiphinema*, are associated with P. *communis* in Kashmir valley. (Table 4.4).

(vii) Pomgaranate *(Punica granata)* (Punicaceae)*:* Pomgaranate has been found to be infested by five species of phytonemaodes, belonging to five different orders. Earliest records of nematode species, *viz. Tylencholaimellus raski* and *Quinisulcius punici,* affecting pomgaranate, have been given by Jairajpuri & Siddiqi (1963) and Gupta & Uma (1980) respectively. Later, more nematode species like *Boleodorus* sp, *Tylenchorhynchus* sp. and *Tylenchus filiformis* on the *P. granata* have been recorded by Fotedar & Kaul (1982) and Kaul (1998). (See Table 4.4).

(viii) Strawberry *(Fragaria verca)* (Rosaceae)*:* Various parasitic nematodes associated with strawberry in Kashmir Valley are : *Helicotylenchus caroliniensis*, *Longidorus*, *Meloidogyne hapla* and *Xiphinema*. These species have been reported by Fotedar & Kaul (1982) and Wali-ullah *et al.* (1983). (See Table 4.4).

(ix) Walnut *(Juglans regia)* (Juglandaceae)*:* About a dozen of plant-parasitic nematode species, distributed over ten different genera, have been found in Kashmir affecting the Walnut fruit tree. Earliest records of nematode species, *viz Pungentus juglansi* and *Discolaimoides indicum,* have been provided by Mahajan (1972). Further, various genera *Hoplolaimus, Longidorus, Paurodontus, Psilenchus, Tylenchorhynchus Tylenchus* and *Belodorus*, have shown to be damaging *J. regia* tree by Fotedar & Kaul (1982). Recently, more nematodes species have been found damaging this fruit tree by Kaul & Wali-ullah (1989) and Kaul (1998). These include *Hoplolaimus indicus (= (Basirolaimus indicus), Pratylenchus juglansi* and *Tylenchus filiformis*. (See Table 4.4).

(D) Mushroom nematodes

Earliest record of pathogenic nematode of mushroom, *viz. Ditylenchus* sp, infesting *Agaricus bisporus* (Lang) from J.&.K, has been given by Sharma *et al.* (1981). Recently, Munshi *et al.* (2001) reported the occurrence of 3 species of nematode parasites on white-button mushroom, *A.brunnescens* Peck. These myceliphagous nematodes are: *Aphelenchoides bessey* Christie, *A. composticola Franklin* and *Ditylenchus* sp.

TABLE 4.4 . Species Diversity of plant-parasitic nematodes, affecting valuable crops (cereal, vegetable & fruit) of J & K State

Type of Crop (a)	Phytonematode Species (b)
(I) CEREAL / FOOD CROP	
(a) Paddy (*Oryza sativa*)	*Aphelenchus* sp., *Hemicriconemoid* sp., *Hirchmanniella gracilis, H. mucronotus, H. shamimi, Tylechorhynchus oryzae, Tylenchus* sp. and *Xiphinena* sp.
(b) Maize (Zea mays)	*Helicotylenchus erythrinae , H. pteracercusoides, H. steineri, Hoplolaimus sp. Longidorus reneyi, Longidorus sp. Paratylenchus sp., Paurodontus sp., Pratylenchus.*
(c) Wheat (*Triticum aestivum*)	*Anguina tritici and Basiria tritici.*
(II) VEGETABLE CROP	
(a) Brinjal (*Solanum melongena*)	*Basiria kashmirensis, Criconemella sp., Helicotylenchus haki, H. mucronatus, H. trapezoidicaudatus, H. silvactus, Helicotylenchus sp. Hirschmanniella gracilis, Hoplolaimus indicus, Meloidogne incognita, M. javanica, Nothotylenchus basiri,Paravulvus confusus ,Pratylenchus melongena, P. zeae, Tylenchorhynchus haki, T. kashmirensis, T. mashhoodi, Tylenchorhynchus sp., Tylenchus andrassyi, T. filiformis, Tylenchus sp. and Xiphinema sp.*
(b) Cabbage (*Bassica oleracea* var. *capitata*)	*Helicotylenchus indica, H. silvactus, Hoplolaimus indicus, Tylenchorhynchus, bassicaecapitata*)
(c) Cauliflower (*B.o.*var.*botrytis*)	*Basiria kashmirensis, Helicotylenchus mucronotus, Hirschmanniella gracilis, Nothotylenchus basiri, Pratylenchus brachyurus, P.similis, Tylenchorhynchus brassicae, T. kashmirensis and T. mashhoodi.*
(d) Chilli (*Capsicum annuum*)	*Helicotylenchus haki, Hirschmanniella gracilis, Meloidogyne hapla, M. incognita, Pratylenchus barkati, T brassicae, T. kashmirensis, T. meshhoodi, Tylenchus andrassyui and T. filiformis*
(e) French Bean (*Phaseolus vulgaris*)	*Criconemella sp., Helicotylenchus indicus, H. minzi, Helicotylenchus sp., and Xiphinema sp.*
(f) Soyabean (*Glycine max*)	*Lordellonema bashirensis*

Table contd...

(a)	(b)
(g) Kale (*Brassica oleracla acephala*)	*Basiria haki, Helicotylenchus dihystera, H. cavenessi, H.haki, Hirschmanniella gracilis, Hoplolaimus indicus, Nothotylenchus basiri, N. srinagarensis, Pratylenchus zeae Tylenchorhynchus haki, Tylenchus andrassyi and Tylenchus filiformis*
(h) Okra (lady finger) (*Ablemoschus esculentus*)	*B kashmiransis, Helicotylenchus mucronotus, Hoplolaimus indicus, Meloidogyne incognita, Nothotylenchus basiri, Pratylenchus similis, Rototylenchulus reniformis, Tylenchorhynchus brassicae, T. kashmirensis and T. mashoodi*
(i) Onion (*Allium cepa*)	*B. kashmirensis, H. dihystera, Helicotylenchus sp. and Nothotylenchus basiri*
(j) Pea (*Pisum sativum*)	*H. dihystera, P. brachyurus and T. mashhoodi*
(k) Potato (*Solanum tuberosum*)	*Aphelenchoides sp., B. kashmirensis, Globodera rostachiensis, H. indicus, Helicotylenchus sp., Hoplolaimus indicus, M. hapla, M. incognita, P. barkati, P. brachyurus, P. penetrans, P. thornei, P. zeae, T. kashmirensis, T. mashhoodi, Tylenchus andrassyi, T. filiformis, Tylenchus sp. and Xiphinema sp.*
(l) Raddish (*Raphanus sativus*)	*B. kashmirensis, H. dihystera, M. incognita, N. basiri, P.zeae, R. reniformis, T. brassicae*
(m) Spinach (*Spinacea oleracea*)	*H. silvactus, M. incognita, T. filformis*
(n) Tomato (*Lycopersicum esculentum*)	*H. dihystera, H. erythrinae, H. indicus, H. trapezoidicaudatus, Helicotylenchus sp., M. incognita, N. basiri, Pratylenchus penetrans, Scutellonema petersi, T. haki, T. kashmirensis, T. mashhoodi, Tylenchorhynchus irregularis, Tylenchus andrassyi, T. filiformis,Tylenchus sp. and Xiphinema sp.*
(o) Turnip (*Brassica campestris rapa*)	*H.indicus, N.basiri, Paratylenchus sp., P. barketi, P. penetrans, P. similis, P. thornei, R. reniformis, T. kashmirensis.*
(p) Winter Squash (*Curubita maxima*)	*Helicotylenchus sp., M. incognita, Psilenchus sp., R. reniformis and Tylenchorhynchus sp.*
(III) FRUIT CROP	
(a) Apple (*Pyrus malus*)	*B. kashmirensis, Boleodorus sp., Enchodelus macronatus, E. zonatus, H. hazaratbalensis, H. papillocauda, H. caroliniensis, H. indicus, Hoplolaimus sp., Longidorus sp. Pratylenchus valvus, Pungenatus angulatus, Tylenchorhynchus sp., Tylenchus filiformis, Tylenchus sp. and Xiphinema sp.*
(b) Almond (*Prunus amygdalus*)	*Criconemella sp., Longidorus sp., Neolobocriconema laterale, Paratylenchus sp., Pratylenchus pseudopratensis and P.vulnus*

Table contd...

(a)	(b)
(c) Apricot (*Prunus armeniaca*)	*Helicotylenchus jammuensis*
(d) Cherry (*Prunus avium*)	*Aphelenchus sp., Criconemella sp., H. dihystera, N. erythrinae, H. kashmirensis, Meloidogyne hapla, Pratylenchus pseudopratensis, Rotylenchus sp. Tylenchorhynchus kashmirensis, T. mashhoodi, Xiphinema basiri and Xiphinema sp.*
(e) Peach (*Prunus persica*)	*Helicotylenchus hazaratbalensis, H. kashmirensis, Helicotylenchus sp. Tylenchorhynchus pruni and T. filiformis*
(f) Pear (*Prunus communis*)	*Helicotylenchus sp., Longidorus sp. and Xiphinema sp.*
(g) Pomegranate (*Punica granata*)	*Boleodorus sp., Quinisulcius punici, Tylencholaimellus raskii, Tylenchorhynchus sp. and T. filiformis*
(h) Strawberry (*Fragaria vesca*)	*H. caroliniensis, Longidorus sp., M. hapla and Xiphinema sp.*
(i) Walnut (*Juglaus regia*)	*Boleodorus sp., Discolaimoides indicum, Hoplolaimus indicus, Hoplolaimus indicus, Hoplolaimus sp., Longidorus sp., Paratylenchus juglansi, Paurodontus sp., Psilenehus sp., Pungenatus juglansi, Tylenchorhynchus sp., T. filfiform and Tylenchus sp.*

TABLE 4.5. An up-to-date catalogue of plant-parasitic nematodes of Jammu and Kashmir, and Ladakh.

Sr. No. (a)	Nematodes species (b)	Host plant /Crop species (c)	Distribution (d)	Author(s) reported / Reference(s) (e)
1.	*Aerotylenchus safroni*- Fotedar & Handoo	*Crocus sativus*	Kashmir	Fotedar & Handoo (1977)Fotedar & Kaul (1982)
2.	*Anguina tritici*	*Triticum aestivum*	Kashmir Ladakh	Shiekh (1979), Fotedar & Kaul (1982) Zaki & Mir (1994) and Zaki (1999)
3.	*Aphelenchoides* sp.	*Solanum tuberosum*	Kashmir	Mahajan (1971), Zaki *et al.* (1991)
4.	*Aphelenchus* sp.	*Prunus avium, Oryza sativa*	Kashmir	Fotedar & Kaul (1982)
5.	*Basiria haki* Fotedar &- Mahajan	*Brassica oleracea* var. *acephala*	Kashmir	Fotedar & Mahajan (1973)
6.	*Basiria kashmirensis* Jairajpuri	*Abelmoschus esculentus, Allium cepa B.o var. botrytis, Pyrus malus, Raphanus*	Kashmir	Mahajan (1971), Jairajpuri (1965), Fotedar & Kaul (1982)

Table contd...

(a)	(b)	(c)	(d)	(e)
		sativum, Solanum melongena, S. tuberosum		
7.	*Basiria tritici*	*Triticum aestivum*	Kashmir	Fotedar & Mahajan (1973)
8.	*Boleodorus cynodoni* Fotedar & Mahajan	*Cyanodon dactylon*	Jammu	Fotedar & Mahajan (1972)
9.	*Boleodorus* sp.	*Juglans regia, Punica granata, Pyrus malus*	Jammu Kashmir	Fotedar & Kaul (1982)
10.	*Cephalenchus rotundus*	Unidentified Plant species	Kashmir	Siddiqi & Khan (1983)
11.	*Chronoplectus ovalvatus* n.gen.et sp. *(Undescribed)*	Roots of unidentified gross	Kashmir	Mahajan(1971)
12.	*Criconemella sp.* (=*Macropothonia*) (=*Criconemoides*)	*Phaseolus vulgaris Prunus amygdalus Prunus avium, S. melogena*	Kashmir	Fotedar & Kaul (1982)
13.	*Discloaimoides indicum* Mahajan	*Juglans regia*	Kashmir	Mahajan (1972)
14.	*Eminesia oranatus* n.gen et. n.sp. (undescribed)	Roots of unidentified grass	Kashmir	Mahajan (1971)
15.	*Euchodelus macronatus* (deMan)	*P. malus*	Kashmir	Jairajpuri & Loof (1967), Mahajan (1971)
16.	*Euchodelus zonatus*	*P. malus*	Kashmir	Jalraljpurl & Loof (1967)
17.	*Globodera rostochiensis*	*A. cepa, B.o. var. acephale & L. esculentum, Pisum sativum, R. sativus and P. avium*	Kashmir	Kaul (1978), Wali-Ullah (1992), Wali-ullah & Kaul (1997)
18.	*Helicotylenchus caroliniensis Sher*	*Fragaria vesca, M. sylvestris*	Kashmir	Kaul (1984), Kaul (1998)
19.	*H. carenessi sher*	*B.o.* var. *acephala*	Kashmir	Zaki *et al.* (1997), Wali-ullah (1992) and Kaul(1998)

Table contd...

(a)	(b)	(c)	(d)	(e)
20.	*H. dihystera* (Cobb)	*A. cepa, B.o.* var. *acephala, L. esculentum., P. avium, P. sativum*	Kashmir	Wali-ullah & Kaul (1997), Kaul (1998)
21.	*H. erythrinae* (Zimme-rmann)	*P. avium, L. aesculentum, Zea mays*	J & K	Kaul & Fotedar (1987) Wali-ullah (1992), Bhat *et al.* (1994)
22.	*H. graminophilus* Fotedar & Mahajan	Unidentified grass species	Kashmir	Fotedar & Mahajan (1974)
23.	*H. haki* Fotedar & Mahajan	*B.o. acephala, Capsicum annuum, S. melongena*	Kashmir	Fotedar & Mahajan (1974) Zaki *et al.* (1991), Kaul (1998)
24.	*H. hazaratbalensis* Fotedar & Handoo	*Prunus persica, P. malus*	Kashmir	Fotedar & Handoo (1974), Fotedar & Kaul (1982)
25.	*Helicotylenchus indicus* Siddiqui	*B.o. capitata, B. campestris rapa, L. esculentum, S. tuberosum, Phaseolus vulgaris*	Kashmir	Zaki *et al.* (1991), Wali-ullah & Kaul (1997)
26.	*H. jammuensis* n.sp (Undescribed)	*Prunus armeniaca*	Jmmu	Mahajan (1971)
27.	*H. kashmirensis* Fotedar & Mahajan	*Prunus persica, P. avium*	Kashmir	Fotedar & Mahajan (1974), Fotedar & Kaul (1982)
28.	*H. minzi*	*Phaseolus sp.*	Kashmir	Zaki *et al.* (1991)
29.	*H. mucronotus Siddiqui*	*A. esculentus, B.o. botrytis,*	Kashmir	Kaul (1998)
30.	*H. papilliocauda* Kaul	*S. melongena*	Kashmir	Kaul (1987)
31.	*H. pteracercusoides* Fotedar & Kaul	*M. sylvestris (=P. malus) Zea mays*	Kashmir	Fotedar & Kaul (1982, 85)
32.	*H. silvactus* Lal & Khan	*B.o. capitata, S. melongena, Spinacea oleracea*	Kashmir	Kaul (1998)
33.	*H. trapezoidicaudatus* Fotedar & Kaul	*L. eseulentum, S. melongena*	Kashmir	Fotedar & Kaul (1983), Zaki *et al.* (1991, 93)
34.	*H. steineri* Fotedar & Mahajan	*Z. mays*		Fotedar & Mahajan (1974)

Table contd...

(a)	(b)	(c)	(d)	(e)
35.	*Helicotylenchus* spp.	*A. sativum, Cucurbita maxima, L.esculentum, Phaseolus vulgaris, P. persica, Pyrus communis, S. melongena*	Kashmir	Fotedar & Kaul (1982), Zaki *et al.* (1991, 93)
36.	*Hemicriconemoides magniferae* Siddiqi	*Cedrus deodera*	Kashmir	Khurma & Uma (1988)
37.	*Hemicriconemoides* sp.	*Oryza sativa*	Kashmir	Fotedar & Kaul (1982)
38.	*Heterodera avenae*	Host Plant unknown	Kashmir	Zaki (1999)
39.	*H. trifolii*	Unidentified medow grass	Kashmir	Swarup et al. (1964)
40.	*Hirschmanniella gracilis* (deMan)	*B.o. acephala, B.o. botrytis, C. annuum, S. melongena*	Kashmir	Wali-Ullah (1992) Wali –Ullah *et al.* (1991)
41.	*Hirschmanniella mucronota* Das	*O. sativa*	Kashmir	Mahajan (1971), Wali-ullah (1991) Luc & Goodey (1963)
42.	*Hirschmanniella oryzae* (Soltwedel)	*O. oryzae*	Kashmir	Mathur & Prasad (1971), Mahajan (1971), Wali-ullah *et al.* (1991)
43.	*H. shamimi* Ahmad	*O. oryzae*	Kashmir	Wali-ullah *et al.* (1991)
44.	*Hoplolaimus indicus* Shamshi (=*Basirolaimus indicus* Sher)	*A. esculentus, B.o. acephala, B.o. var. capitatcu, Juglans regia, P. malus, S. melongena, S. tuberosum*	Kashmir	Kaul (1984), Kaul (1998), Wali-ullah (1992)
45.	*Hoplolaimus* sp.	*J. regia, P. malus, Z. mays*	Kashmir	Mahajan (1970), Fotedar & Kaul (1982)
46.	*Longidorus reneyii*	*Z. mays*	Kashmir	Raina (1966)
47.	*Longidorus* spp.	*Fragaria indica, M. sylvestris, P. amygdalus, Z. mays, J.regia, P. communis, Euronymus japonica*	Kashmir	Fotedar & Kaul (1982), Wali-ullah *et al.* (1993), Kaul (1984)
47a.	*Lordellonema bashirensis* Akhter *et al.*	*Glycinemax*	Kashmir	Akhter *et al.* (1996)

Table contd...

(a)	(b)	(c)	(d)	(e)
48.	*Meloidogyne hapla* Chitwood	*E. vesca, S. tuberosum, P avium, C. annuum*	Kashmir	Krishna & Prasad (1986), Wali-ullah *et al.* (1993), Zaki *et al.* (1991)
49.	*M. incognita* Chitwood	*A. esculentum, C. annuum, C. maxima, R, sativus, S. oleracea, S. melongena, S. tuberosum*	Kashmir	Wali-ullah (1992), Kaul (1998), Bhat & Kaul. (1994)
50.	*M. javanica* (Trenb)	*S. melongena*	Kashmir	Wali-ullah (1992)
51.	*Neolobocriconema laterale* (Mehta & Raski (=*Criconema laterale*)	*P. amygdalus*	Kashmir	Khan & Siddiqui (1963), Mehta & Raski (1971)
52.	*Nothotylenchus basiri* Khan	*A. esculentus, A. cepa, B.o.* acephala, *B.o. botrytis, B. rapa, L. esculentum, R. sativus, S. melongena*	Kashmir	Wali-ullah (1992), Kaul (1998)
53.	*N. srinagarensis* Fotedar & Mahajan	B.o. acephala	Kashmir	Fotedar & Mahajan (1972), Zaki *et al.* (1991)
54.	*Paratylenchus juglansi* Kaul & Wali-ullah	*Juglans regia*	Kashmir	Kaul & Wali-ullah (1989)
55.	*Paratylenchus* sp.	*Brassica campestris, P. amygdalus, Z. mays*	Kashmir	Fotedar & Kaul (1982)
55a.	*Paravulvus confuses* Akhter *et al.*	*S.melongena*	Kashmir	Akhter *et al.*(1996)
56.	*Paurodontus* sp.	*E. japonica, J. regia, Z. mays*	Kashmir	Fotedar & Kaul (1982)
57.	*Pratylenchus barkati* Das & Sultana	*S. tuberosum, B.rapa, c.annuum*	Kashmir	Wali-ullah & Bhat (1990), Kaul (1998)
58.	*Pratylenchus brachyurus* Filipjiv & Stekheven	*B. borytis, P. sativum, S. tuberosum*	Kashmir	Wali-ullah (1990), Wali-ullah (1992)
59.	*P. minimus,* Kaul	*Z. mays*	Kashmir	Kaul (1987)
60.	*P. penetrans* (Cobb.)	*S. tuberosum, B. rapa, S. esculentum*	Kashmir	Wali-ullah (1992), Zaki *et al.* (1993), Kaul (1998)

Table contd...

(a)	(b)	(c)	(d)	(e)
61.	*P. pseudopratensis* Seinhorst	*P. avium, P. amygdalus*	Kashmir	Mahajan (1971), Wali-ullah & Kaul (1997), Kaul (1998)
62.	*P. similis* (Cobb.)	*Abelmoschus esculentum, B.o. botrytis, B.rapa*	Kashmir	Wali-ullah & Bhat (1990), Kaul (1998)
63.	*P. thornei* Sher & Alen	*B. rapa, S. melongena, S. tuberosum.*	Kashmir	Zaki *et al.* (1991,93), Kaul (1990), Wali-ullah (1992)
64.	*P. vulnus*	*M. sylvestris (=P. malus), P. anygdalus*	Kashmir	Kaul (1984,98)
65.	*P. zeae*	*B.o. acephala, S. melengena, S. tuberosum, R. sativus*	Kashmir	Wali-ullah & Bhat (1990), Wali-ullah (1992), Kaul (1998)
66.	*Pratylenchus* sp.	*E. japanica, Mentha arvensis, Z. mays, Z. mays*	Kashmir Ladakh	Fotedar & Kaul (1982), Zaki (1999)
67.	*Psilenchus* sp.	*C. maxima, J. regia, M. arvensis*	Kashmir	Fotedar & Kaul (1982)
68.	*Pungenatus angulatus*	*Pyrus malus*	Kashmir	Jairajpuri & Baqri (1966)
69.	*P. juglansi*	*Juglans regia*	Kashmir	Mahajan (1972)
70.	*Quinisulcius punici*	*Punica granatum*	Kashmir	Gupta & uma (1980)
71.	*Radophholus similis* (Cobb)	*Musa sp.*	Kashmir	Kaul(1998)
72.	*Rototylenchulus reniformis* (Linford & Oliviera)	*A. esculentus, B.rapa, Cucurbita spp.*	Kashmir	Wali-ullah (1992), Kaul (1998)
73.	*Rotylenchus* sp.	*Prunus avium*	Kashmir	Mahajan (1971), Wall-ullah & Kaul(1997)
74.	*Scutellonema brachyurum* (Steiner)	*Magnifera indica*	Kashmir	Mahajan (1971)
75.	*S. petersi*	*L. esculentus*	Kashmir	Mahajan (1971)
76.	*Tylenchorhynchus brassicae* Siddiqui	*A. esculentus, B.o. botrytis, B.o. capitata, C. annuum, R. sativus*	Kashmir	Siddiqi (1963), Wali-ullah (1992), Kaul (1998)
77.	*T. haki* Fotedar & Mahajan	*B.o. acephala, L. esculentum S. melo*	Kashmir	Fotedar & Mahajan (1972,73,74)

Table contd...

(a)	(b)	(c)	(d)	(e)
78.	*T. irruglaris*	*L. esculentum*	Jmmu	Bhat *et al.* (1994)
79.	*T. kashmirensis* Fotedar & Mahajan	*A. esculentus, B.o. botrytis, B.rapa, C. annuum, S. melongena, S. tuberosum, L. esculentum, P. avium*	Kashmir	Zaki *et al.* (1991,93), Wali-ullah (1997)
80.	*T. mashhoodi* Siddiqi & Basir	*A. esculentus, C. annuum, P. avium, B.o. botrytis, P. sativum, L. esculentum*	Kashmir	Wali-ullah & Kaul (1997), Wali-ullah (1992)
81.	*T. oryzae*	*O. sativa*	Kashmir	Kaul & Wali-ullah (1995)
82.	*T. pruni*	*Prunus persica*	Kashmir	Gupta & Uma (1981)
83.	*T. rugosus*	*Salix babylonica*	Kashmir	Siddiqi (1963), Fotedar & Kaul (1982)
84.	*Tylenchorhynchus* spp.	*C. maxima, J. regia, M. arvensis, Morus alba., P. granatum, P. malus, S. melongena*	Kashmir	Fotedar & Kaul (1982)
85.	*Tylencholaimellus raskii*	*Punica granatum*	Kashmir	Jairajpuri & Siddiqi (1963)
86.	*Tylenchus andrassyii*	*B. acephala, C. annuum, S. melongena, S. tuberosum, L.esculentum*	Kashmir	Fotedar & Kaul (1984)
87.	*Tylenchus kashmirensis* Mahajan	*Crocus sativus*	Kashmir	Mahajan (1971, 73)
88.	*Tylenchus filiformis* Micotetzky	*A. esculentums, B.o. acephala, B.o. capitata, C. annuum, L.esculentum*	Kashmir	Kaul (1982), Wali-ullah (1992), Kaul (1998)
89.	*Tylenchus* sp.	*E. japonica, J. regia, O. sativa, L. esculentum, S. tuberosum, S. melongena, Z. mays*	Kashmir	Fotedar & Kaul (1982); Zaki *et al.* (1993)
90.	*Xiphinema basiri*	*Prunus avium*	Kashmir	Wali-ullah & Kaul (1997)
91.	*Xiphinema* spp.	*F. vesca, O. sativa, L. esculentum, P. vulgaris, P. communis, P. avium, P. malus, S. melongena, S. tuberosum.*	Kashmir	Fotedar & Kaul (1982), Zaki *et al.* (1991), Kaul (1998)

Table contd...

TABLE 4.5 (contd). Systematic position of plant-parasitic nematode species, cited in table 4.5. (Serial nos. 1-91 & 47a, 55a)

Serial no. (s) in the table	Order	Sub-order	Super family	Family	Total no. of	
					Genus/ Genera	Species
1, 10, 86-89	Tylenchida	Tylenchina	Tylenchoidea	Tylenchidae	03	06
5-7, 67	-do-	-do-	-do-	Psilenchidae	02	04
2	-do-	-do-	-do-	Anguinidae	01	01
18-35, 44-45, 73-75	-do-	-do-	Hoplolaimoidea	Hoplolaimoidae	04	23
40-43, 57-66	-do-	-do-	-do-	Pratylenchidae	02	14
71	-do-	-do-	-do-	Radopholidae	01	01
70, 76-84	-do-	-do-	-do-	Tylenchorhynchidae	02	10
51	-do-	-do-	Criconematoidea	Criconematidae	01	01
12	-do-	-do-	-do-	Madinematidae	01	01
36-37	-do-	-do-	-do-	Hemicyliophoridae	01	02
11	-do-	-do-	Plectoidea	Plectidae	01	01
17, 38-39	Tylenchida	Tylenchina	Heteroderoidea	Heteroderidae	02	03
48-50	-do-	-do-	-do-	Meloidogynidae	01	03
72	-do-	-do-	-do-	Rotylenculidae	01	01
54-55	-do-	-do-	Tylenchulidoidea	Paratylenchidae	01	02
8-9, 52-53	-do-	-do-	Neotylenchoidea	Nothotylenchidae	02	04
56	-do-	-do-	-do-	Paurodontidae	01	01
4	-do-	Aphelenchina	Aphelenchoidea	Aphelenchidae	01	01
3	-do-	-do-	-do-	-do-	01	01
14	-do-	-do-	Rhabditoidea	Rhabdidae	01	01

Table contd...

Serial no. (s) in the table	Order	Sub-order	Super family	Family	Total no. of	
					Genus/ Genera	Species
46-47	Dorylaimida	Dorylaimina	Longidoroidea	Longidoridae	01	02
90-91	-do-	-do-	-do-	Xiphinematidae	01	02
15-16	-do-	-do-	Dorylaimoidea	Dorylaimidae	01	02
13	-do-	-do-	-do-	Discolaimidae	01	01
68-69	-do-	-do-	-do-	Norbidae	01	02
85	-do-	-do-	Leptonochoidea	Tylencholaimelidae	01	01
47a	Dorylaimida	——	——	——	01	01
55a	-do-	——	——	—	01	01
				Total No.	**38**	**93**

Notes:

(i) Synonyms are given in parentheses, against each species/genera in the table.

(ii) For classification, Fotedar & Handoo (1975) and Luc *et al.* (1990), have been followed.

4.3 PARASITIC LEECHES

The parasitic leeches are included in phylum Annelida and class Hirudinea, are slender, having leaf- like body, without setae. They show the presence of large posterior sucker and smaller or pseudosucker at anterior end, surrounding mouth. These are ectoparasites of warm and cold blooded animals, feeding on their blood.

The leech-fauna of J&K, including parasitic ones, have been studied by Bhatia (1930,34,39), Soota (1959) and Chandra (1982). The important ectoparasitic leech Species of this region are: *Glossiphonia complanata* (Linnaeus), *Helobdella stagnalis* (Linnaeus) (Order, Rhynchobdellida); *Erpobdella indicus* Bhatia, *E. octoculata* (Linnaeus) (Order, Pharynagobdellida); *Hirudinaria granulosa (*Savigny*)*, *Theromyzon mathai* Bhatia and *T. sexoculata* (Moore) (Order Gnathobdellida). *G. complanata* is a ectoparasite of fresh water snails like *Lymnea.* The blood sucking leeches, belonging to the genus *Theromyzon,* enter throats and nostrils of water fowl/birds, like geese and ducks. The medicinal leech found in Kashmir is, *viz. Hirudinana granulosa* , also known as the Indian medicinal leech.

CHAPTER 5

PARASITIC CRUSTACEANS (COPEPODS) AND ACARI

5.1 PARASITIC CRUSTACEAN

In Kashmir, the parasitic crustacean (Phylum Arthropoda: Class Crustacea), are limited to subclass Copepoda and order Lernaeopodoida. The Lernaeopodoids are parasitic on fishes and as ectoparasites, these remain attached to the gills and fins of fishes, drawing the nourishment from host's tissues.

Earlier, the description of a new copepod, *viz., Lernaea kashmirensis,* parasitic on cobitid fish, *Naemacheilus kashmirensis,* from Kashmir was provided by Malhotra & Joyti (1972). Later, Boxhall (1976) also described a new Lernaeopodid, *Pseudolamproglena simplex,* parasitic on fresh water fishes of Kashmir. Description of third new parasitic copepod, *viz. Lernaea gambusiae*, infecting fish, *Gambusia affinis* in this region, has been given by Koul & Raina (1989).

5.2 TICKS

Ticks belong to Ixodida (order), under super-order Anactinotrichida, sub-class Acari, Class Arachnida of Phylum Arthropoda. They are blood-sucking ectoparasites of vertebrates. Ticks are characterized by: leathery integument; presence of movable capitulum or gnathosoma, consists of a base (basis capituli) and the mouth-parts having a hypostome, paired 4-segmented palps and a pair of chelicerae; hypostome with rows of recurved teeth; genital and anal opening present on ventral side and the presence of Haller's organ, a pit like sensory organ on the tarsus of the first pair of legs. The hard-bodied ticks are included in the family Ixodidae and show the presence of a dorsal scutum or shield, which is absent in soft-bodied ticks, belonging to the family Argasidae.

5.2.1 Faunistic and systematic survey

The earliest faunistic study on the tick-fauna of Kashmir region, present in the collection of the Indian Museum, has been accomplished by Sharif (1928). Later, Sen (1938) published a check and host-list of some tick Species from Jammu and Kashmir region. Dutta *et al.* (1963) recorded 8 species of ticks, including *Ornithodoros tholazani* var. *crossi* and O. *lahorensis* as vector of human relapsing fever. Besides these, other Species reported by them are: *Hyalomma isaaci, Boophilus microsplus, Boophilus* sp., *Rhipicephalus haemaphysaloides, R. sanguineus* and *Haemaphysalis montgomeri.*

Tick species, *viz. Haemaphysalis cornupuncta* and *H. kashmirensis*, have been recorded and described by Hoogstraal & Verma (1962) and Hoogstraal & McCarthy (1965) respectively from Kashmir region. Dhanda & Kulkarni (1969) and Dhanda & Bhat (1971) made taxonomical and ecological studies on ticks (*H. corpupuncta* and *H. sundrai*), parasitizing sheep and goats, occurring in various localities of Kashmir.

First comprehensive account of tick-fauna of Kashmir, has been provided by Das & Sharma (1970). These authors have surveyed 22 different species, including some new records, belonging to the family Argasidae as well as Ixodidae, parasitizing birds (poultry, pigeons), rabbit, bat, rodents and even man, occurring in diverse localities/regions of districts, *viz*. Srinagar, Baramulla, Budgam and Anantnag of Kashmir Valley. This paper has listed 7 species of soft-bodied ticks, distributed over 3 genera and 16 Species of hard-bodied ticks spread over six genera.

Sharma & Das (1971) recorded for the first time tick (*Boophilus decoloratus*), parasitizing cows at Lagama (Uri) and also from cows in places like Hazratbal, Safakadal, Saidakadal localities of district Srinagar and in district Baramulla and Anantnag.

Dhanda *et al*. (1971) reported tick–*Dermacentor raskemensis,* infesting Yak in Leh, Flagging in Batalik and goat in Poyen region of Ladakh. The earlier recorded Species of ticks of J&K are incorporated in the checklist prepared by Sharma & Das (1972).

Rao *et al.* (1973) published a general review-cum-checklist of tick-fauna of J&K State, showing 23 species from different districts, viz. Srinagar, Anantnag, Baramulla (Kashmir Province), Doda, Rajouri and Udhampur (Jammu Province) besides, Ladakh Province, attacking hosts like birds, mongoose, sheep, goat, shrew, rodents, horse, dogs, buffaloes, small borrowing mammal from these Districts/ Provinces. Further, according to these authors the Province-wise break-up of tick species was as: 13 species in Kashmir, 17 species in Jammu and 5 species in Ladakh.

A faunal-list of ticks of Jammu region has been contributed by Sharma & Sharma (1976), incorporated about 6 species of Argasids, belonging to 2 genera and 5 species of Ixodids, pertaining to 2 genera, attacking different vertebrate hosts like poultry, birds, bats, sheep, goat, buffalo, cow, rodents, domestic dogs, from Poonch, Jammu, Rajouri and Bhadarwa region.

Dhanda *et al.* (1979) collected nymphs of tick (*Nosomma monstrosum*) from *Mus saxicola* and adults from Buffaloes in Jammu for the first time. Srivastava & Wattal (1981) surveyed ticks in Gulmarg and Tangmarg areas of Kashmir, and recorded four specimens of nymphal stages of *Haemaphysalis bispinosa* from *Rattus rattus*, and of *Boophilus microplus* from cows and buffaloes.

Survey of ticks for Crimean-Congo haemorrhagic fever virus activity in J&K was conducted by Kaul *et al.* (1990), reported 16 species of ticks covering 7 genera as ectoparasites of small and large mammals from Rajouri, Poonch and Udhampur. The newly recorded species by them include *Haemaphysalis intermedia, Hyalomna detritum* and *H. dromendarii*. Recently, Mir *et al*. (1993) showed the prevalence of tick (*Argas persicus*), infesting indigenous fowls of Kashmir.

5.2.2 Species diversity and host range

After scanning number of research papers, including valuable monograph and other works on tick parasites of this region, it was established that a total of about 47 species of ticks, distributed over 10 genera are known to exist in Jammu, Kashmir and Ladakh region. Of these species, 10 species are

Argasids (soft-bodied ticks), distributed over 3 genera and 37 Species are Ixodid (hard-bodied ticks), spread over 7 genera. The total number of tick species known to occur in three Provinces of this State is as 15, 29 and 13 in Jammu, Kashmir and Ladakh respectively. The total number of species under various genera (10), belonging to family Argasidae and Ixodidae of J&K State is as: *Argasids–Argas* (5 spp.), *Ornithodoros* (4), *Otobius* (1); Ixodids–*Boophilus* (4 spp.), *Dermacentor* (3), *Haemaphysalis* (11), *Hyalomma* (8), *Ixodes* (6), *Nosomma* (1) and *Rhipicephalus* (4). (See Table: 5.1 pp. 96-98).

In Jammu and Kashmir, of the total 5 species of ticks, belonging to the genus *Argas*, 3 species have been found to infest birds/poultry and 2 species parasitize bat. In addition to these hosts, *Argas persicus* has sheep, goat and cattle as common hosts, both in Jammu as well as Kashmir region. Genus *Ornithodoros* included four different Species, attacking sheep and goat. Apart from these hosts, 3 Species of this genus have been reported from cattle and 1 Species each from Dog and Rodent (*Rattus rattus*). *Otobius megnini* was recorded from Srinagar, attacking sheep, goat and cattle (Table 5.1).

Amongst Ixodids, *Boophilus* has four Species from J&K State, infesting cattle, of these species 2 species, *viz. Boophilus annulatus* and *B. microplus*, have been found to infest sheep and goat of this region. *Dermacentor* of this area, has 3 Species, all infesting not only the small/burrowing mammals but also sheep and goat. The majority of the species i.e. 11, under the genus *Haemaphysalis* have been reported to attack wide host groups like lizard, birds, small burrowing mammals, cattles, sheep/goat and beast of burden showing, 1, 1, 9, 6, 8, 2 Species respectively, in various districts of Jammu and Kashmir Province, *viz.* Poonch, Rajouri Baramulla, Srinagar, Doda, Udhampur and Anantnag (Table 5.1).

Genus *Ixodes*, represented by 6 Species, found to parasitize sheep and goats of Jammu and Kashmir region. Besides these hosts, species of genus Ixodes have also been reported from cattle, small mammals, dogs and man, with 4 , 3, 2 and 1 species respectively, occurring in wide localities of various districts of Jammu and Kashmir. 8 Species of ticks, pertaining to genus *Hyalomma* are reported from various Districts of J&K. Of these species, 6 Species have been found on sheep and goat, followed by 5 species, affecting cattles, 2 Species each parasitizing small mammals and beast of burden and 1 Species parasitic on dog. (Table 5.1).

Nosomma monostrosum has been found to attack *Mus saxicola* and Buffaloes in Jammu region. *Rhipicephalus* of this region includes 4 Species, reported from Jammu, Kashmir and Ladakh region, with wide host range such as small/borrowing mammals, dogs, cattle, sheep, beast of burden and man, having 4, 4, 2, 4, 1, 1 Species of this genus respectively (Table 5.1).

5.2.3 Economic importance of ticks

The ticks are responsible for causing injury and wounds from bites to domestic animals, poultry and other mammals. The wounds caused by ticks, if the capitulum is torn off in forcible detachment from the hosts, are likely to become infected and result in inflammed sores or extensive ulcers. When ticks are present in large number, these cause serious or even fatal anaemia in cattle or horses. Ticks have also been found to cause paralysis in domestic animals due to toxins secreted by specific cells in the salivary glands. Apart from causing different diseases in domestic animals, ticks also act as vector of viral, bacterial, spirochaetal, rickettsian and protozoan diseases of humans and domestic animals. Different tick-borne diseases, reported from J&K State are briefly highlighted under the following lines:

5.2.3.1. Ticks of veterinary importance

Argas species are considered to be economically important tick Species in this region, parasitizing not only the poultry but also domestic animals (sheep, goat and cattles). Tick-paralysis in fowl and sheep through *Argas persicus* and *Ornithodoros lahorensis* in Kashmir, has for the fist time been reported by Das & Sharma (1970a). *A. persicus* was earlier recorded by Rao & Kalra (1949), present in most fowl houses, examined by these authors.

Another important Argasid tick, *viz. Otobius megnine* (spinose ear-tick) reported from Srinagar, infesting cattle, sheep and goat, is known to choke the ear of these animals, with tick individuals, wax and other debris, leading to rupture of ear drum and also to secondary infection.

Dermatosis disease, characterized by inflammation, itching, swelling and ulceration of site of bite or skin ulceration and lesion resulting from improper or partial removal of tick mouth, found in domestic animals of Kashmir, is due to ticks, *viz. Ornithodoros lahorensis, Otobius megnini and Hyalomma anatolicum* (Das & Sharma, 1970).

The Hard-bodied ticks (Ixodidae) of Jammu and Kashmir, are represented by important genera, *viz. Boophilus, Dermacentor, Haemaphysalis, Hyalomma, Ixodes, Nosomma* and *Rhipicephalus*. Of these tick genera, *Haemaphysalis* is causing maximum number of species i.e., 11. The species of this tick genus, have wide spectrum of hosts, including small/burrowing mammals, birds, cattle, lizard, dogs and other domestic mammals. The second abundantly found genus of this region is *Hyalomma*, with 8 species. These are large tough and hardy ticks, ectoparasitic on domestic animals, small burrouing mammals and beast of burden (horses and camel) of this State.

The genus *Ixodes*, with 6 Species, occurring in wide localities of J&K on diverse hosts. These ticks have small size, easily overlooked when searching a host, feeding catle, sheep, dogs, small mammals and human in this region. *Ixodes ricinus* also known as the European castor-bean tick is an important tick feeding in its all stages on cattle, sheep and dogs.

The genera, *viz. Boophilus* and *Rhipicephalus*, each represented by 4 species in this region. *Boophilus microplus*, *B. decoloratus*, *B. annulatus*, parasitizing large mammals especially cattles. *Rhipicephalus* species are distributed not only in various localities of Jammu and Kashmir, but also in Ladakh region. *R. sanguineus* (the brown dog-tick) most frequently feed on carnivores, mammals and man.

Genus *Dermacentor*, including three Species from J&K, are ornate and bear eyes. These are not host specific. *D. raskamensis* shows wide occurrence and hosts in Ladakh region as: Yak in Leh; Flagging in Batalik and goat in Poyen. Besides, this species of tick parasitizes cattle (Buffalo, goat and sheep) in Kashmir Valley. Genus *Nosomma* having one species from J&K State, *viz. N. monostrosum,* has been recorded from Jammu, attacking buffaloes and small mammal (*Mus saxicola*).

(A) ***Spirochaetosis and Anaplasmosis:*** Avian spirochaetosis disease caused by a spirochaete–*Borrelia anserina*, transmitted by *Argas persicus*, has been reported to render considerable damage to poultry in Kashmir (Das & Sharma, 1970).

other tick-borne bacterial diseases reported from Kashmir are: avian anaplasmosis, caused by rickettsia like parlicle, transmitted by tick (*Argas*) in poultry and bovine anaplasmosis, caused by *Anaplasma marginale*, transmitted by tick, *viz. Boophilus microplus* in cattle and cross bred cow in Srinagar district (Das & Sharma, 1970; Shaw, 1971 and Shaw, 1980). The disease, anaplasmosis was thought to be by previous workers due to protozoan. However, it is now regarded as rickettsian bacterium, belonging to the family Anaplasmataceae.

(B) ***Babesiosis and Theileriosis (Protozoan Diseases):*** Babesiosis disease also known as red-water fever is caused by Haemosporidian protozoan, *viz. Babesia bovis* (= *B. argentina*), has been reported from Kashmir Valley in Jersy bulls and in Australian Jersy cows by Das & Shaw (1968) and Shaw (1980). These authors have shown that babesiosis disease is transmitted by tick, *Boophilus microplus* in this region. For details of Babesiosis, see pages 3-4.

Another Haemosporidian (*Theileria parva*) causes the deadly East-west fever or Theileriosis (bovine), has been found in cross-bred Jersy cows and, sheep and goats of Kashmir, transmitted by tick Species: *Boophilus annulatus* and *Hyalomma anatolicum*. Theileriosis disease of domestic animals in Kashmir has earlier been investigated by Das & Sharma (1970), Shaw & Das (1970) and Shaw (1971). For details of theileriosis disease, see page 4.

5.2.3.2. Ticks of medical importance

(A) ***Tick-typhus:*** The tick-borne typhus is known as spotted-fever (Kettle, 1995), occurring in every continent and zoogeographical region. The disease is caused by Rickettsia. The rickettsiae are transmitted transtadially and transovarially in the tick vector. They cause benign infections of non-human and severe disease in humans which may prove fatal. There is belief that whenever Ixodid ticks bite people, there is possibility of human spotted fever (=Rocky Monutain spotted fever).

A case of tick-typhus in Srinagar was reported by Rao (1951). This author found a patient who was in habit of sleeping with dog in his bed and detailed examination of this patient revealed a tick, *viz. Rhipicephalus sanguineus*, sticking inside his ear and another unattached found by him between his toes. The similar type of ticks were also found on dog and collected from sofa of the patient's room.

(B) ***Relapsing fever:*** An outbreak of relapsing fever (tick-borne) that took place amongst troops in Jammu, was for the first time reported by Kaul (1949). He reported more than hundred cases of this disease, carried by ticks and caused by spirochaetes, *viz. Borrelia duttoni*. Latter, Rao & Kalra (1949), during their course of investigation on the role of ticks in transmission of this disease established that tick, *viz. Ornithodoros crossi* was a vector of a form of relapsing fever in Kashmir. The infection of this disease was described due to the bite of tick and not through coxal fluid or faeces. Tick itself is an effective and important reservoir of infection.

The period between 1948-49, also showed the high-incidence of disease from Army camps in Kashmir. An appreciable density of tick (*O. tholozani* var. *crossi*) was shown to influence the incidence of relapsing fever, starting from foothills of Jammu province and extended into Ladakh. The breeding season of the vector tick did not influence the incidence of disease. This type of tick-borne relapsing fever, transmitted by *Ornithodores tholozani* showed occurrence from Uzbekistan to Kashmir (cf. Harwood & Maurice,1979). The detailed studies on bionomics of vector tick, relapsing fever disease and epidemiology of this disease in J&K State, have been done by: Rao & Kalra (1950); Agarwal (1951), Kalra & Rao (1951) and Dutta *et al.* (1963).

(C) ***Crimean–Congo haemorrhagic fever (CCHF):*** Crimean-Congo haemorrhagic fever is caused by CCHF virus, belonging to the family Bunyaviridae, has earlier been reported from Western Border districts of J&K, *viz.* Rajouri and Poonch districts, based on the samples collected from general population of the area as long term resident of the area (Mohan Rao, 1986). However, in the later study on CCHF activity in these areas by Rodrigues *et al.* (1986), no CCHF virus incidence in the border regions surveyed, was reported either among human or domestic animals. According to these authors that CCHF virus or closely related virus may be present, however, it was not of major health importance. Recently, Kaul *et al.* (1990) during the course of their thorough survey of ticks for CCHF virus activity in J&K State, have reported tick species, *Hyalomma anatolicum*, as an important biological vector of CCHF disease.

The virus of CCHF is enzootic in many regions of the world, including Palaearctic region, chiefly in steppe, semidesert and foothill biotope, where one or two *Hyalomma* species are the predominant ticks, parasitizing domestic and wild animals. CCHF disease involves fever, severe headache and bleeding from every orifice and venipuncture site. The infection occurs in man through the bite of the infected tick or by crushing ticks in contact with the skin or from shearing tick-infected sheep.

TABLE 5.1. An up-to-date catalogue of tick-fauna (Acarina: Ixodida), with their host spectrum and distributions in various districts of J&K state

S. No.	Family/ scientific name of tick	Host/s	Distribution in districts
I. Family: Argasidae (Soft-bodied Ticks)			
1.	*Argas (Carios) vespertilionis*	*Pipistrellus* vespertilionis (Bat)	P, S.
2.	*Argas (Carios)* sp.	*P. minus* (Bat)	P,
3.	*Argas persicus*	Ca., Go., Po., Sh.	P, B, K.
4.	*Argas reflexus indicus*	Pi.,	S.
5.	*Argas* sp.	Po	P.
6.	*Ornithodoros crossi*	Ca., Go., Sh	K
7.	*O. lahorensis*	Ca., Go., Ra., Sh.	B, D, J, K, L., S
8.	*O. savignyi*	Ca., Go., Sh	K
9.	*O. tholazoni*	Do., *Rattus rattus*, Sh.	B, J, S
10.	*Otobius megnini*	Ca., Go., Sh.	S.
II. Family: Ixodidae (Hard-sodied Ticks)			
11.	*Boophilus annulatus*	Ca., Sh.	S.
12.	*B. decoloratus*	Ca.,	A, B, S.
13.	*B. microplus*	Ca, Ga., Sh.	A, B, D, P, R, U

Table contd...

S. No.	Scientific name of tick	Host/s	Distribution in districts
14.	*Boophilus* sp.	Ca	K
15.	*Dermacentor auratus*	Go., Ra.	S.
16.	*D. raskamensis*	Ca, Do., Go., Mo-h, Ro,Sh,Ya and Flagging	Le/L
17.	*Dermacentor* sp.	Go., Sh., Ro.	S
18.	*Haemaphysalis bispinosa*	Ca., Go., Mo., Mu., *R. rattus*	B, J, P, R, S, U.
19.	*H. cornupunctata*	Ca., Go., Dz., Mo-h., Ro.	B, D, J, P, R, U.
20.	*H. howletti*	Bi., Go., Mo., Sh.	R, U
21.	*H. indica*	Mo., Ro., Shr.	U
22.	*H. intermedia*	Ca.	R.
23.	*H. microplus*	Ca., Go., *R. rattus*, Sh.	D, P.
24.	*H. montogomeryi*	Ca., Do., Dz., Ho., Ma., *Apodemus flaricollis, Rattus* sp. *Suncus murinus*	A, B, D, J, R, S, P, U
25.	*H. paraturturis*	Bi., Mo., Ro.	U
26.	*H. kashmircusis*	Ca., Go., Li., Ro., Sh.	A, B, S, U
27.	*H. sulcata*	Go., Sh.	B, D, R,U
28.	*H. sundrai*	Sh.	K
29.	*H. anatolicum*	Ca., Go., Ho., Sh.	B, J, S, U
30.	*H. detritum*	Ca. (Buffalo)	P, R
31.	*H. domendarii*	Cam.	J
32.	*H. excavatum*	Ca., Go., Sh.	S.
33.	*H. isaaci*	Ca. (Buffalo), Go.	J, R, U
34.	*Hyalomma kumari*	Go., Ro	U
35.	*H. marginatum*	Ca., Go.	S
36.	*Hyalomma* sp.	Do., Go., Ro	U
37.	*Ixodes acutitarsus*	Co., Do., Sh.	S
38.	*I. himalayensis*	*Rattus* sp., Ro., Shr.	A, B
39.	*I. ovatus*	Ca., Go., Sh.	P.U
40.	*I. kashmiricus*	Ca., Go., Ma., Sh., Ro	A, B, S
41.	*I. ricinus*	Ca., Do., Sh.	S
42.	*Ixodes sp.*	Go., Mo-h., Sh., Sm..- m	B, D

Table contd...

S. No.	Scientific name of tick	Host/s	Distribution in districts
43.	*Nosomma monostrosum*	Buffaloes and *Mus saxicola*	J, R
44.	*Rhipicephalus haemaphysa-loides*	Ca., Cam Sh., Sm-m, Mo	B, D, J, P, R, U
45.	*R.sanguineus*	Ca, Do., Gor, Shr., Rab., Sh. Ro. Ma, Buffalo	J&K- D, J, U, B, R, S
46.	*R. turanicus*	Do., (Domestic) Go., Sh., Sm-m	B, J, L, R
47.	*Rhipicephalus,* sp.	Sh., Sm.-m	B.R.L

Abbreviations used (Table 5.1)*:* A– Anantnag (district); B- Baramulla, (district); Bi – Bird; Ca. Cattle; Cam- Camel; Co-Cow; D-Doda (district); Do-Dog; Dz.- Dzo; Go. – Goat; Gor. – Goral; Ho. – Horse; J- Jammu; J&K- Jammu and Kashmir State; K – Kashmir; L – Ladakh; Le - Leh; Li - Lizard; Ma. - Mare; Mo.- Mongoose; Mo-h. - Mouse hare; Mu.- Mule; P - Poonch (district); Po - Poultry; R. - Rajuri (district); Ra.- Rat; Rab. - Rabbit; Ro. – Rodent; S - Srinagar (district); Sh. - sheep; Shr.- shrew; Sm.-m - Small mammals; U - Udhampur (district); Ya. - Yak.

Sources*:* Das & Sharma (1970). Dhanda & Bhat (1971); Dhanda & Kulkarni (1969); Dhanda & Rao (1964); Dhanda *et al.* (1979); Dutta *et al.* (1963); Hoogstraal (1962); Hoogstraal and McCarthy (1965); Hoogstraal & Verma (1962); Kaul *et al.* (1978, 90): Mir *et al.* (1993); Rao (1951); Rao *et al.* (1973); Rao & Kalra (1949); Sharma & Das (1971); Sharma & Sharma (1976); Srivastava & Wattal (1981).

TABLE 5.2. Species diversity of ticks on various types of vertebrate hosts in J&K State

Number of species of each genus, parasitizing various vertebrate host group/ s									
S.No./Tick Genus	Lizard	Birds	Bats	Small Burrowing Mammals	Dog	Cattle	Sheep Goat Goral	Best of Burden	Man
		(i)	(ii)	(iii)	(iv)	(v)	(vi)	(vii)	(viii)
1. *Argas*	-	3	2	-	-	1	1	-	1
2. *Ornithodoros*	-	-	-	1	1	3	4	-	-
3. *Otobius*	-	-	-	-	-	1	-	-	-
4. *Boophilus*	-	-	-	-	-	4	2	-	-
5. *Dermacentor*	-	-	-	3	-	1	3	1	-
6. *Haemaphys alia*	1	1	-	9	-	6	8	3	-
7. *Hyalomma*	-	-	-	2	1	5	6	2	-
8. *Ixodes*	-	-	-	3	2	4	6	-	1
9. *Nosomma*	-	-	-	1	-	1	-	-	-
10. *Rhipicephalus*	-	-	-	4	4	2	4	1	1
Total No. of species	01	04	02	23	08	28	34	07	03

Table contd....

Table contd....

(i) = domestic fowl, chicken, poultry, *Gallus gallus*, hens, cocks, pigeon and birds (general)
(ii) = vespertilionid bats
(iii) = rodents, Insectivores, mongoose, rabbit, etc.
(iv) = dog
(v) = cow, buffaloes, yak, dzo, bull, flagging
(vi) = sheep, goat and goral
(vii) = horse, mare, mule and camel
(viii) = human

Serial no. 1, 2 & 3 = soft-ticks (Argasidae)
Serial no. 4 to 10 = hard-ticks (Ixodidae)
* indicates *Ixodes* collected from clothing of man, reported by Pomevantza (1948) *cf.* Hoogstraal (1970) & Das and Sharma (1970)
** indicates, according to Rao (1951).

5.3 PARASITIC MITES

The parasitic mites are small in size and their size ranges from 1–3mm. Mites parasitize not only animals, poultry/birds but also humans. The body of a mite is typically composed of an anterior gnathosoma or capitulum and posterior region, known as idiosoma. The gnathosoma shows laterally the presence of palps and more medially the chelicerae. The mouth-parts are adapted for piercing and sucking blood of the host. The legs of a typical mite are six-segmented.

The mites belong to the sub-class Acari of class Arachnida under phylum Arthropoda (Invertebrata). The Acari has been further divided into two super orders, *viz.* Anactinotrichida (=Parasitiformes) and Actinotrichida (= Acariformes). The Actinotrichida includes three orders, out of these, two orders are parasitic as: Astigmata and Prostigmata, covering various species of mites from different regions and host of J&K State. The Anactinotrichida has four orders, the only two orders, *viz.* Mesostigmata and Ixodida, have representative species from this State. The former incorporates parasitic mites of this region and latter includes ticks, dealt with here separately, see p. 91)

5.3.1 Mites parasitic on insects (housefly and honey-bees)

Sharma & Sharma (1972) recorded the ectoparasitic mites: *Marcrocheles muscae domesticae*, belonging to the family Macrochelidae (Mesotigmata) and *Microtrobium* sp. under family Trombidiidae (Prostigmata), parasitising houseflies (*Musca domestica*), for the first time in Jammu and Kashmir province. The nymphs and adults of mites prey upon the eggs and first stage larvae of houseflies. The ectoparasitic mite on honeybees (*Apis* spp.), *viz. Varroa jacobsoni*, has been reported by Punjabi & Saraf (1969) from Kashmir and by Abrol & Kotwal (1996) from Jammu. This mite is a blood-sucking parasite of *Apis cerana*, *A. mellifera*, causing varroatosis. The wounds caused by *V. jacobsoni* invite microorganisms, leading to infections. Also, *Varroa* is responsible for transmitting bacteria and viruses to honeybees, which cause septicemia and paralysis in them.

Other mites, attacking honeybees in this region belong to genus *Tropilaelaps,* (Mesostigmata: Laelapidae), feed on the haemolymph of larvae and pupae of the honeybees and are considered to be most destructive ectoparasites of bees in J&K State. *Tropilaelaps clareae,* infesting *Apis mellifera* in

Kashmir, has been studied by Shah & Shah (1991) and *T. koenigerum* reported from subtropical areas of Jammu by Abrol & Putatunda (1995). They found that this species of mite was feeding on four species of honeybees, *viz. Apis cerana, A. dorsata, A. florae* and *A. mellifera.*

Parasitic mite of honeybees, *viz. Acarapis woodi* (Prostigmata: Scutacaridae), is responsible for causing disease, known as Isle of Wright disease in *A. cerana* and *A. mellifera* of Kashmir region, infesting not only workers, but also drones as well as queen bees (Shah, 1996). This mite harbours tracheae of bees, it feeds on the haemolymphs of adult bees by penetrating the tracheal wall, leading to passing of bacteria in haemolymph, secretes toxic substances, also clogs air passages mechanically and kills the host.

Recently, Putatunda & Abrol (2003) gave a list of 17 spp. of mites, parasitic on honeybees of Jammu and Kashmir. The newly recorded species i.e.,other than those mentioned above are: *Caryoglyphus lactis, Delfindo* sp., *Euvarroa sinhai, Hypopus* sp., *Kuzinia evae, Kuzinia* sp., *Neocypholaelaps indica, Neocypholaelaps* sp., *Schulzea* sp., *Sennertia* sp., *Tyrophagus longior, T. tropicus, Varroa sinhai* and *Vidia* sp.

5.3.2. Mites parasitic on small mammals, poultry birds, sheep and man

In J & K State, parasitic mites of vertebrates (bat, small burrowing mammals, birds, sheep and man), are belonging to the orders, *viz*. Astigmata, Prostigmata and Mesostigmata. The order Astigmata includes families of parasitic mites, well-defined, showing weakly sclerotized body, slow moving and causing mange and scab disease in animals. The order Prostigmata covers heterogeneous acarines. The family Trombiculidae, under this order is parasitic in the larval stages on vertebrates and some are as vector of scrub typhus. The order Mesostigmata has important families: Dermanyssidae and Macronyssidae, as parasites of Poultry-birds and Laelapidae, as ectoparasite of small mammals in various localities and regions of Jammu, Kashmir and Ladakh (Table :5.3, page 102).

(A) ***Mites parasitic on small mammals:*** Rao *et al.* (1973), during the course of his detailed survey on haematophagous arthropods of Western-Himalaya, reported 10 species of Mesostigmatid mites from Jammu, Kashmir and Ladakh Himalayan regions. These included the existence of 5 spp. of mites from Ladakh, 3 spp. occurring in Kashmir and 6 spp. in Jammu, parasitizing rats, shrews, mice, bat and other small mammals of these regions.

In the recent years, Srivastava & Wattal (1981) also studied parasitic mites of small burrowing mammals from pastures of Tangmarg and Gulmarg localities of Kashmir. They reported 3 spp. of trombiculid mites, belonging to genus *Leptotrombium*, parasitizing *Alticola roylei montosa* from Gulmarg and 4 spp. of mesostigmatid mites, under families, *viz.* Dermanyssidae and Laelapidae as haematophagous ectoparasites of small mammals: *A. r. montosa, Bandicota bengalensis, Rattus rattus* and *R. r. gangutrianus,* existing in the pastures of Tangmarg and Gulmarg

(B) ***Poultry bird mites:*** The earliest study on the abundance of a mite of veterinary importance, *viz. Ornithonyssus bursa* (tropical-poultry-mite) was conducted by Sharma (1972), parasitizing the poultry birds in Kashmir. *O. bursa* occurs either on birds or in nest. Mir *et al.* (1993) investigated the prevalence of ectoparasitic mites in indigenous fowl in Kashmir. These include mange-causing mites *(Knemidocoptes gallinae* and *K. mutans)* and a red-chicken mite or roost mite *(Dermanyssus gallinae)* (Table 5.3).

The *Knemidokoptes* spp. are also spelled as *Knemidocoptes* spp. or *Cnemidocoptes* spp. (*vide* Harwood & Maurice, 1979). *K. gallinae* is a depluming mite, attacks the skin at the bases of feathers of poultry birds. There is intense itching due to this and birds are impelled to pluck feathers. *K. mutans* is also known as the scaly-leg mite, microscopic burrowing mite, a attacking the legs of domestic chickens, causing lifting of scales and swollen condition of shanks, with deformity and also encrustations.

The Roost mite *(D. gallinae),* during day time hide in crevices of poultry houses and roosts, however, attacks birds during night. The heavy parasitization of this mite causes mortality in newly hatched chickens and the adult became weak due to loss of blood and susceptible to diseases. Another poultry mite of Kashmir, *viz. Ornithonysscis bursa,* measuring 1mm in length. This parasite has not only been found on bodies of the birds but also in their nests, in different parts of the world. This species is both a day as well as night time feeder. *O. bursa* is also known to attack humans, causing only temporary irritating effect, because the mite cannot survive on host other than birds

(C) ***Ovine mites:*** Recently, Sharma *et al.* (1996) reported the infestation of parasitic mite, *viz. Psoroptes ovis* (= *P. communis var. ovis*) in Kashmir Marino sheep, in Ganderbal and Kangan localities of Kashmir. *P. ovis* is of great economic importance, known to cause highly contagious psorptic mange disease, responsible for the formation of lesions on any part of the body of sheep. The adult mites puncture the epidermis and feed on lymph and tissue fluids, resulting in the formation of scab. This condition leads to loss of fleece and markedly reducing wool production

(D) ***Mite of medical importance:*** The earliest record of a mite of medical importance, *Leptotrombidium deliense (=Tromsicula deliense),* showing wide distribution in Kashmir and Ladakh region, has been given by Kalra & Rao (1951a). These authors also confirmed the cases of Scrub typhus disease transmitted by *L. deliense* in this region. Scrub typhus is also known as tsutsugamushi fever, caused by a Rickettsian (*Rickettsia tsutsugamushi),* Scrub typhus is an acture febril disease, transmitted by mite (trombiculid). When infected mite feed a person, ulcer like eschar develop at the site of feeding besides, the disease shows fever, severe headach, rash and lymphadenopathy. In untreated cases, the mortality may exceed 30%. Disease can be cured with antibiotic (tetracycline). Scrub typhus is zoonotic in nature, where mite, *Leptotrombidium* and wild rats play important role.

An up-to date checklist (Table 5.3 pp. 102-103), pertaining to parasitic mites of vertebrate depicts, a total of 21 spp., known to occur in different region and localities of Jammu,Kashmir and Ladakh. In addition to these honeybees, a total of 20 Spp. are parasitic on invertebrates (housefly and honeybees *Apis* spp.) in these regions. Number of Species of mites, parasitic on various host groups in J & K, is given as under:

Host		Invertebrates		Vertebrates			
	Housefly	Honeybees	Bat	Rodents and other small mammals	Birds (Poultry)	Sheep	Man
Total Number of species	02	18	01	14	04	01	01

TABLE 5.3. Parasitic mites (Class: Arachnida; subclass Acari) on vertebrates (bat, burrowing mammals, birds, sheep and man) of J and K state

Classification & Scientific name of parasitic mite	Hosts	Locality/ies	Reference/s
Super-order I: Actinotrichida- (=Acariformes)			
Order 1: Astigmata			
Family A: Knemidokoptidae			
1. *Knemidokoptes gallinae* (=*Cnemidokoptes gallinae)*	Domestic fowl (indigenous)	Kashmir	Mir *et al.* (1993)
2. *Knemidokoptes mutans* (=*Cnemidokoptes mutans)*	-do-	-do-	-do-
Family B: Psoroptidae			
3. *Psoroptes ovis* (=*P. comminis var. ovis)*	Merino-Sheep	Kashmir, Kangan, Ganderbal	Sharma *et al.* (1996)
Order 2: Prostigmata			
Family A: Trombiculidae			
4. *Leptotrombidium deliense* (=*Trombicula deliense)*	Rodents, Human	Kashmir	Kalra & Rao (1951)
5. *L. irregulare*	*Alticola roylei montosa*	Gulmarg	Srivastava & Watal (1981)
6. *L. nupestre*	-do-	-do-	-do-
7. *L. titwaleasis*	-do-	-do-	-do-
Super-order II (Anactinotrichida) (= Parasitiformes)			
Order 1: Mesostigmata			
Family A: Dermanyssidae			
8. *Dermanyssus gallinae*	Domestic fowl (indigenous)	Kashmir	Mir *et al.* (1993)
9. *Hirstionyssus sunci*	*Rattus rattus*	Tangmarg/Gulmarg	Srivastava & Wattal (1981)
10. *Hirstionyssus* sp.	Unknown	Ladakh	Rao *et al.* (1973)

Table contd...

Classification & Scientific name of parasitic mite	Host (s)	Locality/ies	Reference/s
11. *Liponyssoides* (= *Allodermanyssus*) sp.	House Mice	Ladakh	-do-
Family B: Macronyssidae			
12. *Ornithonyssus bursa*	Poultry birds	Kashmir	Sharma (1972)
Family C: Laelapidae			
13. *Androlaelaps marshalli*	*R. rattus*	Tangmarg, Gulmarg	Srivastava & Wattal (1981)
14. *Echinolaelaps echidnius*	Rat, Shrew	Doda, Udhampur, Rajouri	Rao et al. (1973)
15. *Eulaelaps stabularis*	Rat, mouse	Ladakh, Tangmarg, Gulmarg	Rao *et al.*(1973), Srivastava (1981)
16. *Laelaps agiliis*	Bat, Bird	Ladakh, Baramulla	Rao *et al.*(1973)
17. *L. algericus*	Small mammals, *R. r. gangutrianus*	Doda, Ladakh,	-do-
18. *L. jugalis*	Rat	Baramulla, Udhampur,	Rao *et al.*(1973) ,Srivastava 1981) Rao *et al.*
19. *L. nuttali*	*A.r. montosa*, *Ban-dicota bengalensis*	Tangmarg Gulmarg,	Srivastava & Wattal (1981)
20. *L. turkestanicus*	Rat	Doda, Baramulla	Rao et al. (1981)
21. *Laelaips* sp.	Rat	Udhampur, Rajouri	-do-

Note: Classification and nomenclature followed in the above table is, as given by Evans (1992) and Kettle (1995).

CHAPTER 6

PARASITIC INSECTS (FLEAS, LICE AND BUGS)

6.1 FLEAS

Fleas belong to order Siphonaptera, are wingless ectoparasites of birds and mammals. The body of adult fleas is laterally flattened and covered with rows of bristles that point backwards. In fleas, the mouth-parts are modified for piercing host skin and sucking their blood, the posterior pair of leg is strikingly adapted for jumping. Fleas possess the ability of transferring from one host species to another, enabling them to transmit disease from animals, mostly rats to humans. The size of flea is 1-6 mm. The females are larger than males of the same species.

6.1.1 Systematic survey and diversity

The earliest records of fleas, parasitizing wide range of mammalian and bird hosts, distributed in different regions of Jammu, Kashmir and Ladakh, are given in outstanding monographs published by: Sharif (1930); Kalra & Rao (1951) and Hopkins & Rothschild (1953, 62, 66, 66a & 71). Iyengar (1973), published a monograph on the Siphonaptera of Indian regions, describing 76 spp. of fleas, including 10 spp. from Kashmir and Ladakh. The detailed surveys and systematic investigations on flea-fauna of this region, have been conducted by Das & Sharma (1970), Kulkarni & Bhat (1972); Rao *et al.* (1973); Smith (1977); Srivastava & Wattal (1981) and Mir *et al.* (1993).

Das & Sharma (1970), reported for the first time, the occurrence of flea (*Ctenocephalides felis orientalis)*, parasitizing goats from various localities of district Srinagar. Kulkarni & Bhat (1972), collected large number of flea , *viz. Neopsylla secura kashmireasis* (= *Neopsylla kashmireasis* Kulkarni & Bhat) from hosts like rodent, shrew like long-tailed mouse *(Apodemes) flavicollis*), house mouse (*Mus musculus),* The Turkestan Rat (*Rattus rattiodes*) and shrew (*Suncus murinus*), occurring in the tract between Phalgam and Kargil at 1500-29000 mt (high-attitude area) of J&K State, (See table 6.1, pp. 107-108).

Rao *et al.* (1973), in his detailed review paper on haematophagous arthropods of Western-Himalayan region, have listed about 26 species of fleas, from various districts of J&K State, parasitizing rodents, shrew, mongoose, dog, goat, sheep, cattle, bat, man, pika, small burrowing mammals, mouse hare and marmot, etc. The Province-wise number of flea species included 14 Species from Jammu, 13 spp. from Kashmir and 17 spp. from Ladakh.

The new taxa and geographical records of some Siphonaptera of small mammals, rodents (6 spp.) of Kashmir, were given by Smith (1977). These included description of a new parasite subspecies, *viz. Amphipsylla phaiomydis imonia,* parasitizing rodent – *Hyperacrius* and taxonomic down grading of *Neopsylla kashmirensis* Kulkarni & Bhat to *N. secura kashmirensis.*

Srivastava & Wattal (1981) reported the existence of three species of fleas, *viz. Nosopsyllus punjabensis, N. fasciatus* and *N. simla,* from *Rattus rattus gangutrianus, Mus musculus* and *Alticola roylei montosa,* occurring in the Tangmarg and Gulmarg areas of Kashmir Valley. Recently, the prevalence of flea, *Echidnophaga gallinacea* in indigenous fowl of Kashmir Valley, has been studied by Mir *et al.* (1993).

A total of 29 species of fleas, distributed over 16 different genera, are now known to occur in different districts and localities of J&K State. The various genera, with total number of Species are: *Acropsylla* (1 sp.), *Amphipsylla* (3 spp.), *Callopsylla* (1), *Ceratopsyllus* (1), *Citellophilus* (2), *Ctenocephalides* (3) *Echidnophaga* (1), *Frontopsylla* (2), *Neopsylla*(3), *Nosopsylla* (4), *Paradoxopsylla* (1), *Pulex* (1), *Stivalius* (2), *Thaumapsylla* (1), *Xenodaeria* (1) and *Xenopsylla* (2 spp.).

The total number of fleas Species, parasitizing various kinds of bird and mammalian hosts including, human, in Jammu, Kashmir and Ladakh, Province is given as under:

Host Type	Total Number of Species		
	In Jammu	In Kashmir	In Ladakh
Birds (Poultry and other aves)	02	03	01
Bat	01	-	-
Small burrowing mammals	11	13	15
Dog	05	05	02
Sheep and Goats	04	04	01
Cattles	04	04	01
Man	01	01	01

6.1.2 Fleas of medical and veterinary importance in J & K State

The main fleas of medical importance found in Jammu, Kashmir and Ladakh region are *Pulex irritans, Xenopsylla astia and X. cheopis,* belonging to the family Pulicidae. *Pulex irritans* also known as the human-flea, parasitizes not only man but also domestic animals. Much irritation is caused to the hosts by the piercing sucking types of month-parts of adult *P. irritans*. *Xenopsylla cheopis* (Oriental rat-flea) is a sole important vector of plague and murine typhus of human in different parts of the world. The incidence of human plague disease has so far not been reported from J&K State. However, Kalra & Rao (1953) reported the existence of murine typhus epidemic in Kashmir valley.

Earlier, due to murine typhus disease in Kashmir, death was reported to be 8-17% besides, *X. cheopis*, lice also played a part in its spread. The high death was not due to the virulence of infection but due to the lack of any medical aid. This disease in Kashmir was confirmed after the trapping of rats, where the cases of murine typhus occurred and rats were found to be parasitized by fleas, *viz. X. cheopis and X. astia.* The individuals of *X. cheopis,* have the habit of easily leaving their host and may transfer to other host of same or of different species. The causative agent of murine typhus disease is a Rickettsian, *viz. Rickettsia mooseri.* It is also known that after the initial infection by rat flea *(X. cheopis),* lice also played a role in transmitting disease from person to person. The murine typhus in man is much milder type of disease.

The important flea Species of veterinary importance of J&K are: *Echidnophaga gallinacea, Nosophyllus nilgiriensis* and *Ceratopsyllus hirudinis,* belonging to the family Ceratophyllidae; *Ctenocephalides canis* and *Ctenocephalides felis* under family Pulicidae. The *E. gallinacea* (the stick-tight flea) of poultry has been found to parasitize the indigenous fowl in Kashmir. Sometimes, it is also called as burrowing fleas, due to impregnated female fleas into the host's skin, causing sever irritation and the skin of the host become swollen. Mir *et al.* (1993) studied the prevalence of *E. gallinacea* in Kashmir fowl, attacking in groups, comb and wattle of the host.

Ctenocephalides canis (dog-flea) and *C. felis* (cat-flea), have been found to attack dog in J & K State.These flea species are not host specific and both these species also parasitize sheep,goat and cattle in this region. *C. canis* is also suitable intermediate host for dog tapeworm,*Diphylidium caninum.* The other fleas of veterinary importance are: *Ceratopsyllus hirudinis,* showing wide distribution in J&K and has been reported to parasitize birds,known to cause serious injury to poultry(hen);*Nosopsyllus nilgiriensis* is responsible for biting cattle in various localities of Jammu,Kashmir and Ladakh Province.

TABLE 6.1 Fleas (Siphonaptera: Insecta) and their vertebrate host diversity in three Provinces of J & K State

S. No.	Fleas species (Scientific name	Province-wise distribution of Host/s of the flea-parasite		
		In Jammu	In Kashmir	In Ladakh
1.	*Acropsylla girshami*	Rodent, Shrew	–	–
2.	*Amphipsylla phaiamydis imonia*	–	*Hyperacrius*	–
3.	*Amphipsylla* sp-I	—	–	Rodent
4.	*Amphipsylla* sp-II	–	–	Rodent
5.	*Callopsylla* sp.	–	–	-do-
6.	*Ceratopsyllus hirudinis*	Birds	Birds	Birds
7.	*Citellophilus* sp-I	–	–	Rodent
8.	*Citellophilus* sp-II	–	–	Marmot
9.	*Ctenocephalides canis*	Dog, Sheep, Goat, Cattle	Dog, Sheep, Goat, Cattle	–
10.	*Ctenocephalides felia felis*	Dog, Goat,	Dog, Cattle,	–

Table contd....

S. No.	Fleas species (Scientific name)	Province-wise distribution of Host/s of the flea-parasite		
		In Jammu	In Kashmir	In Ladakh
		Sheep, Mongoose, Bird, Cattle	Bird, Goat, Mongoose	
11.	*Ctenocephalides felia orientalia*	Dog, Cattle, Sheep and Goat	Cattle, Dog, goat, Sheep	–
12.	*Echidnophaga gallinacea*	–	Kashmiri Fowl	–
13.	*Frontopsylla ambigua*	–	Rodent	Rodent
14.	*Frontropsylla spadix*		–	Rodent
15.	*Neopsylla secura*		Rodent, Shrew	Rodent, Shrew
16.	*Neopsylla secura kashmirensis*	Rodent, Shrew	*Apodemes flavicollis, Mus musculatus, Rattus rattoides, Suncus murinus*	*A. flavicollis, M. musculatus, R. rattoides, S.murinus*
17.	*Neopsylla sp.-I (Setosa group)*		–	Rodent
18.	*Neosopsyllus fasciatus*	–	*Alticola roylei – montosa*	
19.	*Nosopsyllus nilgirionsis*	Cattle and small mammals	Cattle and small mammals	Cattle, & small mammals
20.	*Nosopsyllus punjabensis*	Dog and small mammals	Dog and small mammal, *Rattus rattus ganguntrianus*	Dog, & small mammals
21.	*Nosophyllus simla*	Small mammals	*A. roylei montosa* and small mammals	Small mammals
22.	*Paradoxopsylla custodis*	Rodent	Rodent	Rodent
23.	*Pulex irritants*	Dog, Goat, Sheep, Rodent, and Man	Dog, Goat, Sheep, Rodent, Man	Dog, Goat, Sheep, Rodent, Man
24.	*Stivalius ferinus*	Rodent, Shrew	–	–
25.	*Stivalius shalae*	Small mammals	–	–
26.	*Thaumapsylla breviceps orientalis*	Bat	–	–
27.	*Xenodaeria telios*	–	Rodent, Shrew	–
28.	*Xenopsylla astia*	–	Rat	–
29.	*Xenopsylla cheopis*	Rodent, Shrew, Mongoose	Rat	–

Source: Kalra & Rao (1951); Das & Sharma (1970); Rao *et al.* (1973); Smith (1977); Srivastava & Wattal (1981); Mir *et al.* (1993).

6.2 LICE (SUCKING AND CHEWING)

Lice (sucking and chewing) belong to order Phthiraptera of class Insecta. These show dorsoventrally flattened body, exopterygotes, wingless and as obligatory ectoprasites of birds and mammals. The lice have been grouped into two sub-orders, *viz.* Anoplura (=Siphunculata), including sucking-lice and Mallophaga, covering chewing- lice. In Anoplurans, the mouth- parts are retracted in narrow head, specialized for blood – feeding of mammals mainly. The size of sucking- lice ranges from 5 to 8 mm and remain closely attached to the body of the host. These are transmitted by the body contact. In various Districts and localities of J & K State, the anoupluran Species, belonging to the families, *viz.* Haematopinidae Hoplopleuridae, Linognathidae and Pediculidae, have been encountered.

The Haematopinidae includes genus *Haematopinus* and *Neohaematopinus*, parasitizing sheep, goat, cattle, small burrowing mammals and also poultry. The Hoplopleuridae includes 7 Species of genus *Hoplopleura* and 5 Species pertaining to genus *Polyplax,* parasitic on rodents. 6 Species of the genus *Linognathus* have been found to be parasitic on sheep, goat and cattle, belonging to the family Linognathidae (Table 6.2 pp. 112-113).The medically important family is Pediculidae, including genus *Pediculus,* parasitic on human and monkeys in this region .

In chewing-lice (Mallophaga), the head is as broad as, or border than the thorax, with mandibles pincer-like, present on the ventral side of the head. They spend their entire life- generation and after generation on the host. In Kashmir, the member of the family Menoponidae (Amblycera) under Mallophaga, is reported to parasitize the domestic bird, represented by an important Species, *viz. Menopon gallinae* (Shaft – louse)(Table:6.1). This parasite is about 2 mm in size, found on the thigh, breast, feathers and is harmful to young fowls.

Another important family of Mallophaga from this region is Philopteridae (Ischnocera), incorporating 4 Species(Table: 6.2). These are: the Chicken head-louse *(Cuclotogaster heterrographus),* found on the head and neck region of the host; fluff-louse *(Goniocotes gallinae),* remain attach to down features and causes little irritation; *G. gigas,* largest louse on chicken, occur any where on the body and *Lipeurus caponis* occurs on the underside of the wing and tail features of domestic fowl (indigenous).

6.2.1 Systematic survey and diversity of louse-fauna (sucking and chewing)

The earliest information about the blood-sucking lice from J&K Himalayan region is available through the papers, published by: Kalra & Rao (1949); Ferris (1951); Kalra (1952, 59); Narsimhan (1962); Das & Sharma (1970) and Mishra *et al.* (1973). Rao *et al.* (1973) gave a exhaustive survey report of anoplurans, found in the J&K Himalayan regions. They listed a total of 19 Species of lice from various Districts of J&K State. These included 8 Species from Ladakh, 11 Species from Jammu and 9 Species from Kashmir Valley, parasitizing hosts like rats, mice, goat, shrew, sheep, man, cattle, buffalo, dog and monkey (Table:6.2).

Three Species of lice, parasitic on rodents, mammals and buffaloes, occurring in the pastures of Gulmarg and Tangmarg localities of Kashmir Valley, have been reported by Shrivastava & Wattal (1981). Mishra (1981) published a monograph on the hoplopleurid lice of Indian sub-continent, therein he listed about 7 Species of hoplopleurid lice, parasitizing mammal hosts, belonging to order Insectivora,

Logomorpha and Rodentia. Recently, Mir *et al.* (1993) studied the prevalence of chewing-lice as ectoparasites of indigenous fowl in Kashmir and recorded 1 sp. under family Menoponidae and 4 spp. belonging to family Philopteridae (Table 6.2).

An up-to-date checklist (Table 6.2) of lice (chewing and sucking), parasitic on birds and mammals of Jammu, Kashmir and Ladakh region, indicates a total of 30 Species of lice (Phthirapterans) are existing in this region. These include 25 spp. of sucking-lice and 5 spp. of chewing-lice. Among sucking-lice, 4 spp. belonging to family Haematopinidae, 14 spp. to Hoplopleuridae, 5 spp. to Linognathidae and 2 spp. to family Pediculidae. Various chewing-lice, reported from Kashmir Valley cover 1 sp., *viz. Menopan gallinae,* under family Menoponidae, 4 spp. pertaining to family Philoteridae, as ectoparasites of chicken and domestic fowl (indigenous).

The Species richness and host diversity of louse-fauna of vertebrates (mammals and birds) of J&K State, is given in the following table:

S.No.	Host-type	Sucking-louse	Chewing-louse
1.	Radents/small and burrowing mammals	16	—
2.	Sheep & Goat	05	—
3.	Cattle	03	—
4.	Human	02	—
5.	Dog	01	—
6.	Poultry (domestic fowl)	01	05
7.	Monkey	01	—

The above table shows dominant number of species of sucking-lice as 16, known to parasitize rodents, small and burrowing mammals, including 1 sp.of Haematoponid and 15 of Hoplopleurid. 5 spp. of Anoplurans attack sheep and goats, 3 spp. affect cattle, 2 spp. are parasitic on human (Man) and 1 sp. each is reported to parasitize dog, poultry (duck, hen, cock) and monkey respectively, in various localities and Districts of J & K State. In addition to these, 5 spp. of chewing-lice (Menoponid and Philopterid), have been found to damage poultry (indigenous fowl) in Kashmir.

6.2.2 Louse (sucking) causing and vector of human diseases

The medically important louse family of this region is Pediculidae, having genus *Pediculus*, parasitic on humans and monkeys. The disease caused by *Pediculus humanus* is infection, responsible for causing victim to become highly sensitive, resulting in tissue reactions that is extremely irritating. Sometimes due to louse bites over a long period, the human skin often becomes deeply pigmented. The condition is known as "vagabond's disease". The irritability is not found in the patients suffering from this disease and victim shows immunity. The human body – louse *(P. humanus)* is responsible for transmitting fatal disease of humans known as epidemic typhus, caused by a rickettsian *(Rickettsia prowazekii)*. Besides this, louse-borne human diseases include trench fever (five-day fever) caused by rickettsia, *Rochalimaea quintana* and epidemic relapsing fever, caused by a spirochaete, *Borrelia recurrentis*.

The earliest outbreak of typhus in J&K State was in the year 1894, which followed the next in 1903 and from 1916 onwards the reports were more frequent. The major outbreak was in the year 1903 and extended to 1944, resulting in 2,481 cases and 730 deaths. This outbreak covered whole of Baramula district and also northern part of Anantnag, including Srinagar. Besides Kashmir Valley, outbreaks of typhus disease frequently appeared in hilly tracts of Jammu and Ladakh at refugee camps. The causative agent of this diseases was a rickettsian. The epidemic typhus disease showed high fever, eruptive rashes, purplish spotting of the skin, severe nervous and gastro-intestinal symptoms, occurrence of microscopic nodules around arterioles and great reduction of leucocytes. The transmission of typhus in man results due to contamination of bites or scratches by faeces or from inhalation of dried faeces of lice.

The first documented report, pertaining to typhus outbreak in Banihal (Kashmir) has been provided by Kaul (1944). Later, during the winter of the year 1947-48, there were two outbreaks of typhus, one started in December, 1947 in refugee camp at Baramulla and Srinagar, and other in March, 1948 at village Brari Pathri (Budgam). The symptoms of epidemic typhus of limited outbreaks were of typical epidemic typhus. The cause was confirmed to contact and the strains isolated from patients and lice, were similar to *Rictettsia prowazekii* (Kalra & Rao, 1949).

The spread of epidemic typhus in earlier days in J&K State, was prevented by de-lousing. However, in the year 1958 and 1960, Srinagar area was surveyed, with the aim to determine the insecticide susceptibility status of the "Hariganiwan strain", "Wasan-Anderwan" and "Malaru" – local strains of the body louse *(Pediculus humanus corporis)*. It was Narsimhan (1962) who determined the status of insecticide susceptibility of local strains of body-louse. He found that with the withdrawal of control measures at Hariganiwan, the local strain reverted to normal susceptibility to DDT and BHC and retained its susceptibility to an indigenous make of insecticide, *viz.* Pyrethrin. It was also observed by him that the strains at Wosan-Anderwan and Malaru, showed susceptibility to all the three insecticides.

Descriptions of an epidemic of typhus, which occurred for the first time in Ladakh in July 1961, was given by Patheja (1963). It was confirmed in Tibetan refugees at Leh. The infection was brought in by refugees, when they migrated to Leh in 1960 from eastern Tibet. It was also established on the basis of available data that the epidemic was louse-borne.

TABLE 6.2. Checklist of lice (sucking and chewing) (Order: Phthiraptera) of vertebrates (Mammals and birds) of Jammu and Kashmir

S. No. (a)	Systematic position/ scientific name of Louse (b)	Host/s (c)	Locality/ies (d)
	(I) Anopleura (= Siphunculata) (Sucking Lice)		
	Family 1: *Haematopinidae*		
(1)	*Haematopinus eurysternus*	Duck, hens, cocks, goat, cattle, musk shrew	Kmr.
(2)	*Haematopinus tuberculatus*	Cattle (calf)	Gul.,Udh.,Raj,
(3)	*Neohaematophinus palearctus*	*Marmota caudata*	Ldk
(4)	*Neohaematopinus petaurista*	*Petaurista petaurista* albiventer (Asian giant, flying squirrel)	J&K
	Family 2: Hoplopleuridae		
(5)	*Hoplopleura acanthopus*	*Mus musculus*	Gul., Tan
(6)	*H. captiosa*	Rat, Shrew, *Mus* sp., *Apodema flavicollis*	Sgr, Ana, Udh., Bar.
(7)	*H. himalayana*	Field Mouse, *A. flavicollis*	J&K
(8)	*H. kondana*	*Millardia mettada*	J&K
(9)	*H. phaiomydis*	*Phaiomydis* sp.	Ldk.
(10)	*H. pacifica*	Rat, Mice	Sgr., Ant., Ldk.Bar., Udh.
(11)	*H. ramgarh*	*Mus.* Sp.	Udh.
(12)	*H. sicata*	Rodents, *Rattus fulvescens*	Agr., Ana., Ldk.Bar., J&K
(13)	*Hoplopleura* sp.	Rat	Udh.
(14)	*Polyplax indica*	*Glumda* sp.	Dod., Udh., Raj.
(15)	*P. rectinata*	Shrew,Mice – *Crocidura* sp., *Suncus murinus*	Dod., Bar., Ach.
(16)	*P. servata*	Mouse	Ldk., Bar
(17)	*P. stephensi*	Rat	Udh., Raj.
(18)	*P. spinulosa*	Rat, *M. musculus, Bendicot bengalensis*	Gul., Dod., Udh.

Table contd...

(a)	(b)	(c)	(d)
		Family 3: Linognathidae	
(19)	*Linognathus apicanus*	Goat	Ldk.,Bar.,Udh.
(20)	*L.ovillus*	Sheep & Goat	Sgr.
(21)	*L. setosus*	Dog	Dod., Udh.
(22)	*L. stenopsis*	Sheep & Goat	Ldk.
(23)	*L. vituli*	Cattle	Ant.,Ldk.,Bar.
	Family 4: Pediculidae		
(24)	*Pediculus humanus corporis*	Goat, Man, Monkey	Sgr.
(25)	*Pediculus* sp.	Monkey	Udh.
	Sub-order 2: Mallophaga (Amblycera and Ichnocera) (Chewing lice)		
	Family: 1. Menoponidae (Amblycera)		
(26)	*Menopan gallinae*	Domestic fowl (indigenous)	Kmr.
	Family: 2. Philopteridae (Ichnocera)		
(27)	*Goniocotes gigas*	Domestic fowl (indigenous)	Kmr.
(28)	*G. gallinae*	-do-	Kmr.
(29)	*Cuclotogaster heterographus*	-do-	Kmr.
(30)	*Lipeurus caponis*	-do-	Kmr.

Abbreviations used: Ach. – Achabal; Ana – Anantnag; Bar. – Baramulla; Dod. – Doda; Gul. – Gulmarg; J&K – Jammu & Kashmir; Kmr. – Kashmir; Ldk. – Ladakh; Raj. – Rajouri; Srg. – Srinagar; Tan. – Tangmarg; Udh. – Udhampur.

Source: Narsimhan (1962); Das & Sharma (1970); Mishra (1981); Mishra *et al.*(1973); Mir *et al.* (1993)

6.3 PARASITIC BUGS

A rare bug(Hemiptera), *viz. Psammotettix striatus* (Linn.) has been found to parasitize the high-population of flying fox(bat) (Chiroptera), *Pterops*, which were hanging on the trees: *Plantinus orientalis* (Chinar), *Cedrella toona*(toon) and *Eucalyptus lanceolatus* (eucalyptus) in Poonch Valley, Jammu (Sharma,1973).

In J & K State, two species of bed bugs,belonging to the family Cimicidae (Hemiptera), *viz.Cimex hemiperus* and *C. lectularius,* have been reported. Bed bugs are cosmopolitan, nocturnal, usually found in large numbers and are known to suck blood of animals, poultry (chicken) and man.They can be extremely annoying,persons who are victim of biting of bed bugs,suffer loss of sleep,sores from infected bites and iron/haemoglobin deficiency.

Rao(1973), reported the existence of *C. hemipterus* and *C. lectularius* from Ladakh region ,*C.hemipterus* from Jammu(Udhampur and Rajouri),from indoor places.The prevalence of bed bug,*C.lectularius* as an ectoparasite of indigenous fowls in Kashmir Valley, has been reported by Mir *et al.*(1992).

CHAPTER 7

OTHER PARASITIC INSECTS (MOSQUITOES AND FLIES) (DIPTERA)

7.1 MOSQUITOES

Mosquitoes belong to insect order Diptera, sub-order Nematocera and family Culicidae. In J&K State, the various Species of the genus, *viz. Anopheles* belong to sub-family Anophelinae and genera like *Aedes, Culex* and *Culiseta,* came under sub-family Culicinae (Table 7.1). The adult mosquitoes are parasitic, whereas the larval stages (immature stages) are aquatic. The adult female mosquitoes are blood-sucking parasites. The blood meal is required by many Species of mosquito for their eggs maturation or to increase the ratio of egg production. The adult male mosquitoes feed on vegetation/plant material.

The adult mosquito, size ranges from 3–6 mm, having long forwardly directed proboscis, equal in lengths to head, thorax combined, wings show scales along the veins and forewing fringe, and second vein not arched but running parallel to vein I and II.

The adult of living Anophelinae and Culicinae can be recognized by the posture they attain when resting on a surface. The adult *Anopheles* rest with proboscis, head, thorax and abdomen in one straight line, making an acute angle with the surface. The Culicine adults rest with their body angled and abdomen directed backwards the surface on which it is resting. *Aedes* and *Anopheles* can also be recognized by colour and spots. The colour of *Aedes* mosquito is greyish black, with white stripes. In case of *Anopheles,* the wings have dark spots.

The mosquitoes are potential biological vectors of important human, domestic animal and poultry diseases in many parts of the world, including India and J & K State. In many parts of India, excluding Kashmir, *Aedes* Species are involved in the transmission of dengue disease to humans. Anopheline mosquitoes are medically important Species in being the main vector of human malaria disease in different districts and localities of J & K State. The main *Anopheles* mosquitoes known to transmit malaria disease in Kashmir and Jammu Province are: *Anopheles (Cilia) culicifacies* and *A. (C.) fluviatilis* respectively (See table 7.1 page 117).

7.1.1 Systematic survey, diversity and species richness

The earliest studies on mosquito-fauna of J & K State, including Ladakh region, have been done by: Barraud (1924, 24a), Christopher (1933) and Puri (1948). Bhat & Kulkarni (1971), for the first time reported the occurrence of *Culex (Neoculex) hortensis* from Ladakh region.

Rao *et al.* (1973), during the course of their extensive surveys, conducted between the year 1966–70, in various districts and provinces of this State, established the existence of about 30 Species of mosquito. They showed the district-wise break of various Species as: Doda, 5 spp.; Udhampur, 17 spp.; Rajouri, 3 spp.; Srinagar and Anantnag, 3 spp. Each; Baramulla, 6 spp. and Ladakh (province), 9 spp.

Recently, Bhat & Kulkarni (1983) gave the records of 27 spp. of mosquito, collected from various regions and localities of Jammu, Kashmir and Ladakh, with ecological notes. These included 7 Species, belonging to genus *Anopheles,* 19 spp., pertaining to genus *Culex* and 1 sp. of the genus *Culiseta.*

Now, a total of 32 Species of mosquito are known to occur in various localities, Districts and Provinces of this State. These Species, with classification and distribution range, are presented in Table 7.1. The listed Species cover: 18 spp. from Jammu (5 spp. of genus *Anopheles,* 4 spp. of *Aedes* and 9 spp. of *Culex);* 12 spp. From various districts of Kashmir (8 spp. of genus *Anopheles,* 1 sp. of *Finalaya* and 3 sp. of *Culex).* Apart from these, 9 spp. of mosquito, reported so far from Ladakh province, belonging to genus *Aedes* (3 spp.), to *Culex* (5 spp.) and to *Culiseta* (1 sp.).

Two Species of mosquitoes, *viz. Anopheles culicifacies* and *A. fluviatilis,* have been confirmed to be as vectors of human malaria disease in Jammu and Kashmir. *A. culicifacies* has been reported from Udhampur and other parts of Jammu, whereas the existence of *A. fluviatilis* has been recorded from Jammu (Udhampur) as well as Kashmir (Karnah, Handwara and Uri), see Table:7.1.

7.1.2 Human malaria in Kashmir

In Kashmir, the incidence of human malaria, existed in many villages at higher-altitudes (above 1,524 mts). This was for the first time shown by Jacob (1950), during his survey between April and October. Later, in sixties, the high incidence of fever among the inhabitants of some villages in and around Uri, Tetwal and Keran region of district Baramulla of Kashmir, led National Malaria Eradication Programme (NMEP) Organisation, to collect blood samples from these fever cases and also to conduct entomological survey for Anopheline as vector of malaria disease in Kashmir Valley.

During August–September (1963), C.P. Nair of National Institute of Communicable Diseases, Delhi, conducted an epidemiological study in district Baramulla to establish local transmission of malaria, incriminate vector and to determine the degree of endemicity. The various findings of Nair (1973) are briefly given as under:

The malaria survey incorporated about 30 villages (15 in Karnah, 8 in Uri and 7 in Handwara) of district Baramulla of Kashmir. The percentage of person either suffering from fever or gave very recent history fever, were 10.2% in Handwara, 11% in Karnah and 12.5% in Uri. The most widely prevalent malarial parasite Species was found to be as *Plasmodium vivax* (88.4%) and *P. faliciparum* (11.6%). The inhabitants of the villages, situated at an elevation from 1,524 to 1,8289 mts, showed the cases of *P. faliciparum.* Two individuals were also found to have mixed infection. Apart from this, in one case, it was *P. vivax* and *P. faliciparum* in the other *P. vivax, P. malariae.*

The percentage of heavy infection to total infection was shown to be 74.7%. The gametocytes carrier among the positive ranged from 77.4% in case of *P. falciparum* and 88.9% in case of *P. vivax.* In both the infections, the gametocyte density was also shown to be high.

During the course of entomological survey for anopheline as vector of malaria, *Anopheles fluviatilis* was identified as a vector of malaria, distributed near an elevation of 8,129 mts. The mosquito Species was collected at Pringle Haridal village in Karnah and one individual of *A. fluviatilis,* out of 56 dissected showed the presence of sporozoites of malarial parasites. The density of this mosquito was .33 to 5.3 per man hour. Apart from this, *A. culicifacies* and *A. leucosphyrus* were also present in various villages surveyed, however, in very low density at high-altitude.

Keeping in mind, the various epidemiological findings, Nair (1973) has inferred that an unstable endemic malaria existed in Baramulla district of Kashmir Valley, at all the elevation from 1,676–2,134 mts, in different localities surveyed by the author and transmission of this disease was seasonal, with prevalence mostly during warmer months. This malaria had characteristics of that of subtropical zone malaria.

TABLE 7.1 Checklist and distribution of Mosquitoes (Culicidae:Diptera) of Jammu, Kashmir and Ladakh regions

S. No.	Sub-family/tribe/mosquito Species	Distribution in various districts and localities of J&K State
	I. Sub-family Anophelinae	
1.	*Anopheles (Anopheles) gigas-simlensis*	Bar.
2.	**A. (Cilia) culicifacies*	Udh., Kar., Jmu.
3.	**A. (Cilia) fluviatilis*	Udh, Kar., Han., Uri., Jmu.
4.	*A. (Cilia) aculates*	Dod., Kar., Han., Uri, Udh., Bar.
5.	*A. (Cilia) subpictus*	Jmu.
6.	*A. (Cilia) splendidus*	Han.
7.	*Anopheles hyrcanas*	Kar.
8.	*A. leucophyrus*	Kar.
9.	*Anopheles* sp.	Sgr., Udh.
	II.Sub-family Culicinae **A. Tribe: Aedini**	
10	*Aedes (Aediomorphus) vexans*	Ldk.
11.	*Aedes (Finlaya) elsiae*	Raj.
12.	*A. (F.) ereophilus*	Ldk.
13.	*A. (F.) sintoni*	Ldk.
14.	*A. (Stegomyia) albopictus*	Udh.
15.	*A. (S.) vittatus*	Raj.
16.	*A. (S.) walbus*	Udh.
17.	*Finlaya versicolor*	Yus.
	B. Tribe: Culicini	
18.	*Culex (Culex) barraudi*	Udh.
19.	*C. (C.) bitaeniorhynchus*	Udh.

Table contd....

S. No.	Sub-family/tribe/mosquito Species	Distribution in various districts and localities of J&K State
20.	*C. (C.) fuscocephalus*	Udh.
21.	*C. (C.) mimeticus*	Dod., Sgr., Ldk.
22.	*C. (C.) pseudovishnui*	Ldk.
23.	*C. (C.) theileri*	Karg.
24.	*Culex (C.) vagans*	Ldk., Ana., Bar.
25.	*C. (C.) vishnui*	Jmu., Udh.
26.	*Culex (Culiciomyia) viridiventer*	Ana., Pah., Udh.
27.	*Culex (C.) pallidothorax*	Jmu.
28.	*Culex (Neoculex) hortensis*	Ldk.
29.	*Culex pipiens fatigans*	Bar., Dod., Raj., Udh.
30.	*Culex P. quinquefasciatus*	Ban., Bar., Jmu.
31.	*Culex* sp.	Ldk.
	C. Tribe: Culisetini	
32.	*Culiseta (Allotheobaldia) longiareolata*	Karg.

* indicates as vectors of human malaria disease

Key to Abbreviations cited: Ana.–Anantnag; Ban.–Bandipore; Jmu.–Jammu; Han.–Handwara; Kar.–Karnah; Karg.–Kargil; Ldk.–Ladakh; Pah.–Pahalgam; Raj.–Rajouri; Srg.–Srinagar; Udh.–Udhampur; Yus.–Yusmarg;

Source: Barraud (1924); Bhat & Kulkarni (1971); Nair (1973); Rao *et al.* (1973) and Bhat & Kulkarni (1983).

Note: Classification and Nomenclature followed, as given by Kattle (1995) and Rao *et al.* (1973) respectively.

7.2. SANDFLIES

Sandflies have been placed under order Diptera of class Insecta, belonging to the family Psychodidae, as well defined group of sandflies species, are sometimes accorded family rank as the Phlebotomidae, however, more often regarded as sub-family Phlebotominae. Sandflies resemble tiny moth because of their hairy bodies and moth like appearance. Sandflies are small in size and usually yellowish or buff coloured. The female sandfly is haematophagous and nocturnal in habit, whereas the males are non-parasitic, feeding on moisture.

Sandflies are vector of various types of leishmaniasis (visceral and cutaneous) disease in humans. In Jammu and Kashmir region, visceral leishmaniasis and cutaneous leishmaniasis, caused by protozoans, *viz., Leishmania donovani* and *L. tropica* respectively, have been reported during early fifties.

Jacob & Kalra (1951) and Mitra & Roy (1953) (cf. Kumar *et al.,* 1984) were first research workers to detect few cases of visceral as well as cutaneous leishmaniasis in south-eastern part of Kashmir.

The visceral leishmaniasis is also known as Kalazar. The synonyms of Kalazar are Black disease, Dum-dum fever and Tropical splenomegaly. This disease is mainly characterized by weight loss, hepato-splenomegaly and anaemia. Death can occur in untreated cases of visceral leishmaniasis. In cutaneous leishmaniasis disease of human, the patients develop nodular and ulcerative skin lesions. This infection, if treated properly may heal completely in a few months to few years.

7.2.1 Systematic survey and species richness

Earlier surveys on sandfly-fauna of Jammu and Kashmir regions have been made by: Sinton (1932, 1933, 33a), Theodar (1948), Jacob & Kalra (1951), Mitra & Roy (1953) and Mitra (1959). A total of 9 different spp. of sandflies were observed by Jacob & Kalra (1951), occurring in different localities of Jammu, Kashmir and Ladakh region. They collected 7 spp. from Jammu, 5 spp. from Kashmir and 3 spp. from Ladakh, including one new Species. In Riasi locality of Jammu, they found great variety of sandfly, covering four different Species. Sandfly, *viz. Phlebotonus minutus* was found to be most abundant in Jammu (Riasi). In Kashmir, *P. major* was reported to be uniformly distributed and *P. chinensis,* occurring in villages in Zanaskar range (Jacob & Kalra,1951).

Rao *et al.* (1973) in his detailed paper on haematophagous insects of Western Himalayan region, reported the existence of only one species of sandfly, *viz. Sergentomyia baghdadis* from Rajouri district of J&K, distributed at 500 to 750 mts. They have reported the mode of collection as indoor. More information regarding the existence of sandflies in J&K, with 8 spp. from Jammu and 5 spp. from Kashmir, and 4 spp. as common for both areas, has been provided by Lewis (1978).

Recently, Kumar *et al.* (1984), during the course of his extensive survey for the sandflies in the Jammu area in district Kathua and Udhampur in the month of August 1982, collected 5 spp. both from indoor situation (houses, cattle shed, etc.) and as well as outdoor (under stones, culverts, tree-holes and burrows). These species included *Sergentomyia (Parrotomyia) baghdadis, S. (Neophlebotomus) zeylanica, Phlebotomus (Phlebotomus) papatasi, Sergentomyia (Grassomyia) indica* and *S. (Parrotomyia) kauli.* According to these authors, the highly abundant species was *S. (P.) baghdadis* (46.86%) and least prevalent *S. (P.) kauli* (0.62%) of the total collection made. The sandflies, *viz., S. zeylanica* and *S. kauli* were encountered by these authors for the first time from Jammu, further according to them a total of 11 spp. of sandflies are known to occur in Jammu & Kashmir.

It is evident from the Table 7.2 (page120) that a total of 14 spp. of sandflies, belonging to family Psychodidae (sub-family Phlebotominae) of insect order Diptera, are known to occur in J&K State. About 12 spp. have been recorded from Jammu, 5 spp. from Kashmir and 2 spp. from Ladakh. The species known to exist in Ladakh region are, *viz. Phlebotomus (Adlerius) chinensis chinensis* and *P. (Paraphlebotomus) sergenti.* The species common for both Jammu and Kashmir areas include *P. (A.) c. chinensis, P. (Larroussius) major major* and *P. (P.) papatasi. P (A.) c. chinensis,* showed wide distribution in many localities/areas of Jammu, Kashmir and Ladakh Province of this State.

7.2.2 Kalazar (leishmansiasis) in Jammu and Kashmir

The evidence of human visceral leishmaniasis in Kashmir and Jammu Province, has been given by Jacob & Kalra (1951) and Khan *et al.* (1975) respectively. The earlier definite cases of Kalazar patients in this

State, have been detected in local inhabitants of Kashmir Valley, who had acquired infection inside State. Jacob & Kalra (1951) confirmed this disease in two Kashmiris, one belonging to the district Anantnag and other to district Srinagar, who had never gone outside the State. According to these authors, 12 cases were reported from J&K and these included 7 Gurkhas, 2 Kashmiris and 3 Dogras. The 2 Kashmiri had never been outside State and 3 Dogras (2 from Samba and 1 from Poonch) has also never left the State in their life. The village of Budil was incriminated epidemiologically, as likely endemic area, which had only one species of sandfly i.e. *P. major*, as the most abundant species in whole of Jammu and Kashmir. Further, the authors did not find parasites in tissues of the sandflies. The indigenous cases of visceral leishmaniasis have been reported during 1975 from Jammu region: village Dugan in Kathu district and in Kishtwar region of district Doda.

TABLE 7.2 Checklist of Sandflies (Diptera: Psychodidae) of J&K State

S. No.	Name of Species	Distribution in Province		
		Jammu	Kashmir	Ladakh
1.	*Phlebotomus (Adlerius) chinensis chinensis*	+	+	+
2.	*P. (A.) longiductus*	+	-	-
3.	*P. (Larroussius) major major*	+	+	-
4.	*P. (Paraphlebotomus) sergenti*	-	+	+
5.	*P. (Phlebotomus) papatasi*	+	+	-
6.	*P. minutus* var. *antennatus*	+	-	-
7.	*P. montanus*	-	+	-
8.	*P. squamipleuris*	+	-	-
9.	*Sergentomyia (Grassomyia) indica*	+	-	-
10.	*S. Montana*	+		
11.	*S. (Neophlebotmus) zeylanica*	+	-	-
12.	*S. (Parrotomyia) baghdadis*	+	-	-
13.	*S. (P.) kauli*	+	-	-
14.	*S. (Sergentomyia) punjabensis*	+	-	-

\+ denotes, the existence of sandfly Species in the Province.
\- denotes, not recorded in any locality of the Province

Sources: Jacob & Kalra (1951), Rao *et al.* (1973) and Kumar *et al.* (1983).

7.3 BLACKFLIES

Blackflies or buffalognats are important pests and vectors of veterinary and medical importances, belonging to Order Diptera, Sub-order Nematacera, Family Simuliidae, Subfamily Simuliinae and Tribe Simuliini, characterized by small-sized, robust and hump-backed body. The females are haematophagous and feed on blood of human and animals, whereas the males feed on nectar. Simuliids breed in aquatic habitats. Simuliids are vectors of nematode, *Onchocerca volvulus*, causing river blindness disease in man in Afrotropical and Neotropical regions of the world, and also vectors of *Onchocerca* spp. in domestic animals. Besides, these transmit blood protozoan, *Leucocytozoon* among birds and domestic poultry in different regions of the world.

The earliest records of blackflies from Kashmir, belonging to two Species of genus, *Simulium*, has been given by Edwards (1927). Later, Puri (1932, 32a) described two more Species from this region. Datta (1983,86) studied the blackfly-fauna of Kashmir and recorded few more species from this region.Now, a total of eight species of blackflies are known to occur in various regions of Kashmir. These Species are given as under:

(i) *Simulium (Odagmia) adventicium* Datta; **(ii)** *S.(Eusimulium) aureum* Fries; **(iii)** *S. (Simulium) barraudi* Puri; **(iv)** *S.(S.) chrystophersi* Puri; **(v)** *S. (S.) griseifrons* Brunetti; **(vi)** *S. (Himalayanum) indicum* Becher; **(vii)** *S. (Wilhelmia) mediterraneum* Puri and **(viii)** *S. (Testisimulium) stevansoni* Edwards.

7.4. TABANID, MUSCIDS AND PUPIPARA

Tabanid or horsefly belongs to the family Tabanidae (Diptera),is large-sized and stout-bodied fly, with eyes brilliantly coloured, having zigzag bands, as in *Haematopota*. Horsefly is haematophagous and the mouth-parts well suited to mechanical and biological vector of diseases. Rao *et al.* (1973) reported the existence of horsefly, *viz. Haematopota faciata*, as human pest in Ladakh.

According to Sewell & Roonwal (1965), four species of dipterous flies, belonging to the family Muscidae (Muscoidea), as haematophagous, like blood of animals and man, are known to occur in many areas of Kashmir Himalayan regions. These muscid Species are: *Helina (s.str) cashmirensis*, often haematophagous in man, *Lyperosia (Glossinella) titillans* Bezzi, *Musca (Eumusca) autumnalis* DeGeer, *M. (Viviparomusca) bezzi* Patton & Cragg and *Morellia hortorum cashmiranas* van Emden, found to bite cattle/horses and on cow/horse dung, distributed in various localities like Dal Lake, Ganderbal, Nigeen, Kokernag and high-altitude areas of Gulmarg.

Pupipara Parasites

The dipteran pupipara parasites include ked/louseflies and batfly, belonging to the family Hippoboscidae and Nycteribiidae respectively. These are specialized ectoparasites that live on the body of birds and mammals, and both sexes feed on blood from host. In pupiparid insects, the female flies retain egg inside the uterus, giving nourishment to the developing larva and do not extrude the young until they are ready to pupate.

In J & K, the main Species of Hippoboscid of economic importance is sheep ked, *Melophagus ovinus*, parasitizing sheep and cattle. This parsasite has completely lost its wings and spends entire life-cycle, sheltered in the fleece of its host. Four Species of hippoboscid, are reported to parasitize birds in Jammu and Ladakh region, these include, *Crataerina pallida, Lynchia albipennis, Ornithophila metallica,* and *Pseudolynchia canariensis. P. canariensis* (Pigeon lousefly), attacks nestling birds, causing much irritation to them and also acts as the intermediate host of common pigeon protozoan parasite. (*Haemoproteus columbae*), in other parts of the world.

Hippobosca longipennis and *H.variegata,* have been found to parasitize goat, cattle and dog in Jammu, Kashmir and Ladakh region. Apart from these, lousefly parasites of cattle and domestic animals, goat is also attacked by a hippoboscid, *viz. Lipotena capreoli*, in Doda, Rajouri, Srinagar and Udhampur region. Batfly, *Eucampsida hyrtii orientalis*, belonging to family Nycteribiidae, has been reported to parasitize bat in Udhampur. The batfly is having atrophied wings and remains continously attached to their hosts. The pupae of this parasite are found at the walls of caves, etc., inhabited by bats.

In J & K State, about 9 spp. of pupiparous parasites, have so far been recorded. These cover 8 spp. of louseflies, spread over 7 genera, parasitizing birds, dogs, sheep, goats and cattle, and 1 sp. of batfly parasitic in bat. Of these 9 spp., 7 spp. are from Jammu and 3 spp. each are from Ladakh and Kashmir.

The Pupirara parasites (ked/louseflies) and batfly recorded from different Districts/Provinces of J&K State, are classified as: Order, Diptera; Sub-order, Cyclorrhapa; Series, Schizophara; Section, Pupipara; Superfamily, Muscoidea

S.No.	Family scientific name	Host	Locality(ies)
	Family I : Hippoboscidae (Ked/louseflies)		
1.	*Crataerina pallida*	Bird	Ladakh
2.	*Hippobosca longipennis*	Dog,goat	Ladakh, Baramulla, Udhampur, Rajouri
3.	*Hippobosca variegata*	Cattle	Udhampur
4.	*Lipotena capreoli*	Goat	Srinagar, Doda Udhampur, Rajouri
5.	*Lynchia albipennis*	Bird	Rajouri
6.	*Melophagus ovinus*	Sheep	Srinagar, Ladakh, Cattle Baramulla
7.	*Ornithophila metallica*	Bird	Doda
8.	*Pseudolynchia canariensis*	Bird	Udhampur, Rajouri
	Family II : Nycteribiidae(Batfly)		
9.	*Eucampsida hyrtlii orientalis*	Bat	Udhampur

Source: Sharma & Das (1971) and Rao *et al.* (1973).

7.5 MYIASIS –CAUSING FLIES

In this region, some flies, *viz.*, blowflies/blue-green bottle flies, fleshflies, warblefly, sheep nostrilfly and syrphidfly, belonging to the order Diptera, Sub-order, Cyclorrapha and Group Schizophora, are responsible for causing myiasis in Vertebrates, and also in some Invertebrates. In myiasis, the larvae of the fly feed on host's tissues (healthy or necrotic), fluids(blood or other) or ingested food. When larvae/maggot invade unbroken skin, wound or lesion, it is called as dermal myiasis/traumatic myiasis or creeping myiasis (larvae borrowing beneath the skin). When larvae penetrate the nasal passage, eye-socket or pharyngeal cavity, it is nasopharyngeal myiasis. When the invasion of larvae is in the host's alimentary canal, it is known as gastero-intestinal/entric myiasis. Urinogenital myiasis refers to the entry of larvae into urinogenital system.

Blowflies belong to the family Calliphoridae, are medium-sized flies, cause myiasis in man and animals. In addition to myiasis, these flies play an important role in transmitting pathogenic organisms, carried on the hairs of their legs. Calliphorids find use in treatment of wounds, not responding to medication, in chronic osteomyletis and also in biosurgery, through Maggot therapy and Maggot debridement therapy, in number of hospitals of advanced countries (Baer, 1931 and Sherman & Pechter, 1998).

In Kashmir Valley, four Species of calliphorid are known, these include *Calliphora pattoni* Aubertia, *Gulmargia angustisquma* Rognes, *Lucilia porphyrina* (Walker) and *Pollenia rudis* (Fabricius), reported by White *et al.* (1940) and Rognes, 1972. *Calliphora pattoni*, blue bottle fly, is a larvivorous species, *Lucilia poryphyrina*, green bottle fly, breeds on carcases of birds and animals. These species cause dermal/cutaneous myiasis. *P. rudis* is reported to be parasitic on earthworm, *viz. Allolobophora chlorotica* Savigay.

Fleshflies (Muscoidea: Sarcophagidae) are parasitic during larval stages, showing gray or silvary body. They also act as myiasis-causing agents by invading wounds, responsible for cutaneous furuncular myiasis, characterized by inflammatory nodules or pus pockets. Some sarcophagids are known to penetrate into living tissues, causing severe condition and even death. The hosts of fleshflies are different mammals including man and even insects. Sarcophagid fly larvae have also medical use in Maggot therapy treatment (Gordh & Headrich, (2001).

In Kashmir Valley, 4 spp. of sarcophagids, belonging to genus *Sarcophaga* have been reported by White *et al.* (1940). These include *Sarcophaga dux* Thomson, *S. haematodes* Meigen, *S. khasiensis* Senior-white and *S. melanura* Meigen. The bionomics of first three species is unknown, however, *S. dux* has been shown to cause tissue myiasis in dead animals (squirrel, bovine) and human.

Warbleflies (Muscoidea: Hypodermatidae) are hairy black and yellow-coloured flies, showing bumble bee like appearances, with reddish yellow pile at the end of the abdomen. The females lay their eggs on the hairs of the lower parts of legs or flants of cattle/domestic animals, causing great annoyance. This can lead to a condition known as 'gadding', resulting in weight loss and reduction in milk production. The eggs of Hypodermatids, about 100 or more hatch into spiny larvae which burrow into skin and then for several months they ramble and thereafter, the migrating larvae are found in viscera, abdomen and thorax.

In Ladakh region, Ranga-Rao (1994) showed the prevalence of parasitic infection due to cattle warble fly. *viz. Hypoderma lineatum* on Indian yak, *Bos (Poaphagus) grunniens*. Shahardar *et al.* (1993) reported the existence of goat warble flies, *viz. Hypoderma crossi* in goats of Kashmir, causing skin lesions and loss of hairs in these domestic animals.

Sheep-nostrilfly or sheep/head maggot belongs to family Oestridae (Muscoidae), are moderately large bee-like flies, greyish brown in colour. It is represented in Kashmir by an important species,

Oestrus ovis, parasitizing sheep and goats. The female fly is ovoviviparous which deposits larvae in the nares and long maggots find their way back into nostrils. The larvae cause a profuse discharge of mucous (snotty nose). The larvae develop in the naso-phanyngeal cavities of sheep and goat, causing myiasis. Shahardar *et al*. (1998), for the first time, reported the occurrence of *O. ovis* in the nasal passage and adjoining frontal sinuses of sheep and goats, from the Jammu & Kashmir State.

Syrphidflies or flowerflies belong to the family Syphidae, are free-living, with dark bodies ornamented with brightly colour spots and bands, hovering over flowers. Datta & Chakraborti (1983), during the course of survey on flowerflies of Kashmir, recorded the existence of syrphid fly, *viz*. *Eristalis*, (*Erastalis*) *tenax* (Linnaeus), (Milesinae: Volucellini), from number of localities of Kashmir, *viz*. Gulmarg, Phalgam, Hazratbal, Kabhtarkhana, Rainawari and Lal Mandi.

According to Cheng (1973) the larvae of *E. tenax* are most common myiasis causing species (rectal, gastero-intestinal myiasis) in different parts of the world. The flies lay eggs over cesspools, liquid manur, open privies and waterbodies, contaminated with rotting animals matter. The eggs develops into larvae bearing long-breathing siphon, resembling a tail, hence commonly known as rat-tailed maggot.

CHAPTER 8

ENTOMOPHAGOUS PARASITIC INSECTS

8.1. ENTOMOPHAGOUS INSECTS (PARASITES AND PARASITOIDS)

The entomophagous parasitic insects include parasites and parasitoids, belonging to the order Hymemoptera and Diptera, and involve feeding upon insects. Most insects, parasitic upon other insects are protetean *i.e.*, they are parasitic only in immature stages (larvae) and adults as free living. They devour all or most of the hosts body and then pupate, either within or external to the host. The adult parasites emerge from the pupa and starts the next generation. The adult parasite feed on honeydew, nectar, pollen and many feed on host's body fluid. In general, parasitic insects are distinguished from entomophagous parasitic insect in that their hosts are typically vertebrates, and parasitic as adults and larvae/nymphs; smaller than their hosts, feed upon all stages of their hosts and do not kill their hosts, for example, Siphonaptera (fleas), Anoplura (sucking-lice), Mallophaga (chewing-lice), etc., dealt with separately, (see chapter, 6, pp. 105-113).

Typically insect parasitoids attack one stage of host *i.e.* egg/larva/nymph or pupa. The parasitoid larvae kill their host. The parasitoids are of two types, *viz.* ectoparasitoids and endoparasitoids. In ectoparasitoids, larval stages feed and develop on the external surface of the host's body, whereas in case of endoparasitoids, larval stages feed and develop inside the body of the host. The parasitoids are common among some groups of hymenoptera and diptera.

Now-a-days, entomophagous parasitic insects (parasites/parasitoids) are being successfully utilized for suppression/control of insect pest populations by adopting various methods of applied biological control. Some parasites are non-beneficial, known as hyperparasites, as parasite of parasitoids. The original parasite of insect/pests known as primary parasite and the hyperparasite as secondary parasite. When hyperparasites are present, they play a negative role in biological control programme.

Various entomophagous insects (parasites/parasitoids) known to exist in J & K Himalayan regions, belong to the insect Orders, Superfamilies and Families as: Diptera-Conopoidea (Conopidae), Muscoidea (Muscidae, Sarcophagidae and Tachnidae); Hymenoptera-Ichneumonoidea (Aphidiidae Braconidae Ichneumonidae), Ceraphronoidea (Megaspilidae, Chalcidoidea (Aphelinidae, Chalcididae, Encyrtidae, Eulphidae, Eupelmidae, Eurytomidae, Mymandae, Pteromalidae, Torymidae and Trichogrammatidae) Cynipoidea (Cynipidae). The diversity, host range and economic importance of these parasitic groups are discussed under the following heads:

8.1.1 Dipteran entomophagous insects

In Kashmir Valley, various dipteran parasites reported so far, belong to the Superfamily and family: Conopoidea (Conopidae) and Muscoidea (Muscidae, Sarcophagidae and Tachinidae). The Conopids are also known as tickheaded flies. These are solitary, internal parasites of hymenoptera (bumble bees and wasps), dipteran and also Orthoptera (grasshoppers). The adults of Conopids are free living, however, larvae are parasitic on insect hosts. Females deposits eggs on host's body while in flight, into the abdominal cavity of host. From Kashmir Himalayan region, Brunetti (1920) has reported three Species of Conopids, *viz. Ceria dimidiatipennis, Myopa nigriventris and M. testacea.* The muscids, parasitizing some unknown Species of insects occurring in Dal lake, Ganerbal and Gulmarg localities, reported by Sewell & Roonwal (1965), include *Graphomyia maculata, Phaonia aronitens, P. thomsoni, P. falvomaculata, P. kashmirensis, Pogonomyia fumipennis, Hydrotaea armipes, H. dentips, Mydaea (Myiospila) mediatabunda, agustifrons* and *Muslina stabulanus*. A muscid, *Muslina stabulanus*, has been found to parasitize forest/fruit tree pest, *Lymantria obfuscata* on *Salix* spp. and *Populus* spp. (See Catalogue, p. 138).

A sarcophagid (*Sarcophaga* sp.), commonly known as fleshfly, usually myiasis-causing, however, at times acts as an internal parasite of other insects, is known to occur in this region. *Sarcophaga* sp. has been found to parasitize *L. obfuscata,* infesting *Salix* spp. This dipteran parasite is blackish in colour, with grey thoracic stripes. The larval stages feed on animal/insects and adults feed on sugar containing material. See Catalogue, p.138.

About 6 Species of Tachinids, belonging to 5 genera are reported as endoparasitoids of larval lepidopterans (*L. obfuscata* and *Plutella xylostella*) from Kashmir. Tachinid flies are covered with stout bristles or setae, usually drab coloured. Various Tachinids: *Corcelia* sp., *Compsilura* sp. *Drino,* spp. (2 spp.) and *Exorista rossica,* are known to parasitize larvae of *L. obfuscata* (gypsy moth), serious insect pest of forestry importance in this region. Apart from these, tachinid *(Voria ruralis)* has been reported to attack agricultural pest *P. xylostella* (Diamond back Moth). An insect Species when attacked by the tachinid, is always killed. See Catalogue, p.139.

8.1.2. Hymenopteran entomophagous insects

The hymemopteran parasites/parasitoids, belonging to different families, are discussed under the following heads:

8.1.2.1. Aphelinids (survey, diversity, species richness and as biocontrol agent)

The Aphelinids are small chalcids, belonging to the family Aphelinidae and superfamily Chalcidoidea, assigned to order Hymenoptera. Tk such as fruit and fruit trees (*Prunus* spp., *Pyrus malus),* forest trees (*Populus* spp., *Salix* spp. *Robinia pseudoacaia, Acer caesium, Celtis australis), Dalphne oleoides Verbascum thapsus, Pennisetum flaccidum, Spirea* sp., *Epilobium hirsutum, Viburnum foetans* and *Zinnia elegans*. For details, see Catalogue, p. 138 and Checklist, p.153.

10 different genera, covering 20 spp. of aphids on various host plants are known to be parasitized by different Species of aphelinid parasites and these include *Aphis* (6 spp.) *Brevicoryne* (1), *Drepanosiphum* (1), *Epipemphigus* (1 sp.) *Eriosoma* (1), *Lipaphis* (1), *Macrosiphum* (1), *Myzus* (2), *Pemphigus* (5) and *Rhopalosiphum* (1). 6 spp. belonging to 6 genera of scale insects, attacked by aphelinid parasites in this region are: *Ceroplastodes, Eulecanium coryli Lepidosaphes kirgisica, Parthenolecanium corni, Pulvinaria inconspiqua* and *Quadraspidiotus perniciosus* (San Jose Scale).

The list of various Species of aphelinids occurring in J & K region and their host insect and plant Species, with their distribution and authors reporting, are encapsulated in separate Catalogue, p. 138 and Checklist, p.153.

Aphelinids as biological control agents of insect pests:

The earliest reports of *Aphelinus mali* as an effective parasite of woolly apple aphid, *Eriosona lanigerum*, damaging apple trees in Kashmir valley have been provided by Narayanan (1957) and Rao *et al.* (1971). In India, *A. mali,* in the year 1937, was first released in woolly apple aphid pest infested orchards in Raison (Kulu) Himachal Pradesh for biological control of this pest, in later years, the parasite (*A. mali*) spread to Kashmir and became active in various areas of this region.

In Kashmir valley, for the biological control of San Jose scale, a serious insect pest of apple deciduous fruit trees, the Russain and Chinese strains of an aphelinid parasite, *Aphytis diaspidies*, has been utilized in the year 1964/1965. Later, a Japanese strain of *Aphytis* sp. nr. D*iaspidis* from California (USA) was introduced to Kashmir. The recovery of this parasite was reported from Sumbal and Wadora region. The average percentage of parasitism in these localities during the period August, 1967 to July, 1968 were 68.6% and 61.5% respectively (Rao *et al.,* 1971).

Masoodi *et al.* (1989) released aphelinid, *viz. Aphytis* sp. *proclia* group to control San Jose scale in three apple orchards at Humuhama, Awantipora and Narbal in Kashmir, during the year 1985 and 1986. The activity of this parasite was monitored through sticky traps at Narbal. These authors reported the establishment of parasites in three areas.

The effective parasites of San Jose Scale as *Encarsia perniciosi* (= *Prospaltella parniciosi)* has been found to play a significant role in suppressing this pest in Kashmir orchards. This specific parasite of San Jose scale, has established in many areas, showing high-population density during summer (Masoodi *et al.*, 1989a and Bhagat *et al.* 1992).

Singh *et al.* (1984), for the first time showed the natural parasitization in the eggs of gypsymoth (*L. obfuscata*), a major pest of silvicultural importance, due to parasite, *viz. Encarsia* (= *Prospaltella) kashmirensis* at Fruit Research Station, Shalimar, Srinagar. The parasitization of eggs of *L. obfuscata* was shown to be ranging from 15.78-20.89%.

8.1.2.2 Aphidiids (survey, species richness and their host-aphid and plant diversity)

The aphidiid parasitoids belong to the family Aphidiidae, under super-family Ichneumonidae, however, some authors treat it as a sub-family Aphidiinae of the family Braconidae. Aphidiids are relatively small and fragile wasps, having real joint or articulation below second and third abdominal segment, and reduced venation of forving. All species under this family, are primary and solitary internal parasitoids of aphids (plant-lice). Aphids are one of the important groups of insect pests of agricultural and forestry importance, and of other economically important plants, responsible for causing damage to the host-plants by sucking and also by being vectors of virus diseases.

The female aphidiid parasitoid deposits eggs into the host aphid, in a few days develops into larva and eats out the inside of aphid. It cuts a hole in the bottom of aphid's empty skin and through this glues the carcass to the substratum. The full grown larva of most genera spin a cocoon inside empty skin of the victim. In case of genus, *Praon*, a peculiar tent like cocoon beneath the victim (host aphid) is made. The adult parasitoid emerges by cutting a circular lid. The aphidiid parasitoids are of prime importance in checking the natural population of aphids and are being successfully utilized for biological control of aphid pests in different parts of the world.

Earliest new records and descriptions of new Species, of aphid parasitoids from Jammu and Kashmir region, have been given by: Dharmadhikari & Ramaseshia (1970), Stary & Bhagat (1978), Shuja-uddin (1978*a*) Takada & Rishi (1980) and Bhagat (1981a, 84, 86) Later, number of new Species of aphidiids from this region were described by Bhagat (1981 *c*, 82 *a*, 82*c*) and Shuja-uddin (1970, 74, 78, 83). Studies on aphid-galls and their parasitoids were done by Bhagat (1982). The records and host- range of hyperparasitiods of aphids from Kashmir, have been given by Bhagat (1983). New records, host range and biological notes on aphidiids of Jammu region, have been given by Bhagat & Ahmad (1995).

In Kashmir Valley, detailed investigations on aphidiid parasitoids, controlling aphids, damaging valuable agricultural crops and other economically important plants, have also been carried out. In this direction, observations have been made on the seasonal occurrence of aphids, abundance of parasitoids and their role as natural biological control agents. The various papers encompassing such studies are on: medicinal plants (Bhagat, 1987 *a*); cereals, pulses and vegetables, fodder crops and ornamental plants (Bhagat, 1986 *b*); fruit trees (1986 *a*); trees of forestry and sericulture importance (Bhagat, 1985 *a*). The wide aphidiid parasitoid spectrum of aphids of Kashmir, is having mixed fauna of Palaearctic and Oriental region, and might be useful for exporting of some of the efficient parasitoids to other regions of our country and to other areas of the world for the bio-control of aphid pests.

Aphidiid Species richness and their host-aphid and plant diversity

In Jammu and Kashmir, till date, a total of 74 Species of aphidiids have so far been reported. The various genera, with their total number of species are: *Adialytus* (3 spp.), *Aphidius* (16), *Aeropraon* (1), *Binodoxys* (2), *Diaeretiella* (1), *Ephedrus* (10), *Kashmiria (1), Lepolexis* (2), *Lysaphidius* (1), *Lysiphlebia* (1), *Lysiphlebus* (3), *Monoctonus* (3), *Pauesia* (4), *Praon* (7), *Toxares* (4) and *Trioxys* (15). These parasitoids are known to attack a total of 83 host aphid species, distributed over 44 genera, infesting / damaging more than 111 host plant species, distributed over 73 genera, covering 35 plant families. See Table 8.1, p.130, Catalogue, p.138 and Checklist, p.153.

The various host aphid genera, with their total number of species, attacked by parasitoids are: *Acyrthosiphon* (3 spp.), *Amphicercidus* (1), *Aphis* (9), *Aulocarthum* (1), *Betacallis* (1), *Brachycaudus* (2), *Brevicoryne* (2), *Capitophorus* (2), *Cavariella* (3), *Chaitophorus* (6), *Chactosiphon* (3), *Cinara* (5), *Coloradoa* (1), *Eriosoma* (3), *Eumyzus* (1), *Hyalopterus* (2), *Hypermyzus* (1), *Impatientinum* (1), *Lachnus* (2), *Losomaphis* (2), *Lipaphis* (1), *Macrosiphoniella* (7), *Macrosiphum* (6), *Melanaphis* (2), *Microlophium* (1), *Mindarus* (1), *Myzaphis* (2), *Myzus* (7), *Ovatus* (1), *Phorodon* (1), *Prociphilus* (1), *Periphyllus* (2), *Pterocomma* (1), *Rhopalosiphum* (1), *Schizoneurella* (1), *Semiaphis* (1), *Shivaphis* (1), *Sipha* (3), *Sitobion* (1), *Tetraneura* (1), *Therioaphis* (3), *Tuberlachnus* (1), and *Uroleucon* (4).

The different host plant families (35), incorporating various genera, with total number of Species in case of each genus, damaged by above given aphid genera (44), with 83 Species, parasitized by 16 aphidiid genera (74 spp.), are given as under:–

Family Aceraceae – *Acer* (1 sp.); Apiaceae – *Conium* (1); *Daucus* (1), *Ferula* (1); Asclepiadaceae – *Calotropis* (1); Asteraceae – *Achillea* (1); *Chrysanthemum (3), Cichorium* (1), *Senecio* (1), *Sonchus* (3), *Zinnia* (1); Balsaminaceae – *Impatiens* (1); Berberidaceae – *Berberis* (2); Boraginaceae – *Cyanoglossun* (1); Brassicaceae – *Brassica* (2), *Capsella* (1), *Descuraina* (1), *Sisymbrium* (1); Cannabaceae – *Cannabis* (1); Caprifoliaceae – *Lonicera* (1); Cupressaceae – *Thuja* (1); Dioscoreaceae – *Doscorea* (1); Euphorbiaceae – *Euphorbia* (1); Fabaceae – *Lathyrus* (1), *Lespedeza* (1), *Medicago* (3), *Pisum* (1), *Robinia* (1),

Trifolium (1), *Vicia* (1); Hippocastanaceae – *Aesculus* (1); Fagaceae – *Quercus* (1); Lamiaceae – *Mentha* (3), *Nepeta* (2), *Salvia* (1); Meliaceae – *Melia* (1); Nymphaeaceae – *Nymphaea* (2); Ongraceae – *Epilobium* (1), Pinaceae – *Abies* (1); *Pinus* (3); Poaceae – *Arundo* (1); *Hordeum* (1), *Pennisetum* (1); *Phragmites* (1), *Poa* (1), *Sorghum* (1); *Triticum* (1); Polemoniaceae – *Phlox* (1); Polygonaceae – *Rumex* (2); Punicaceae – *Punica* (1); Ranunculaceae – *Ranunculus* (2); Rosaceae – *Cotoneaster* (1), *Malus* (1) *Prunus* (3); *Rosa* (6), *Rubus* (2), *Spirea* (1); Rubiaceae – *Rubia* (1); Salicaceae – *Populus* (3), *Salix* (7); Scrophulariaceae *Verbascum* (1), *Veronica* (1); Simaroubaceae – *Ailantus* (1); Solanaceae – *Lycopersicon* (1); Ulmaceae – *Celtis* (1); *Ulmus* (2); Urticaeae – *Urtica* (1), See Checklist, p. 153.

The aphidiid parasitoid, *viz. Aphidius*, with highest number of species i.e. 16, has been found to be attacking 42 Species (26 genera) of aphids, infesting 52 spp. (41) of host plants. *Trioxys*, having 15 spp. parasitized 23 spp. (14 genn.) of aphids, infesting 34 spp. (24 genn.) of host plants. *Ephedrus* is represented by 10 spp., attack 16 spp. (13 genn.) of aphids on 16 spp. (13 genn.) of host plants. The next abundant parasitoid, *Praon* (7 spp.) attack 18 spp. (13 genn.) of aphids, infesting 15 spp. (12 genn.) of host plants. *Toxares,* with 4 spp., show 11 different Species of aphid hosts, under 8 genera on 10 spp. (10 genn.) of host plant. *Pauesia*, also having 4 spp., reported to have 4 different species, under 2 genera of aphid host, infesting 4 spp. (3genn.) of host plants. Genus *Adialytus*, *Monoctonus* and *Lysiphlebus*, having 3 spp. each, are parasitic on 3 spp. (3genn.), 4 spp. (3 genn.) and 8 spp. (5 genn.) of aphids infesting, 3 spp. (3 genn.), 4 spp. (3genn.) and 10 spp. (8 genn.) of host plants respectively. *Binodoxys* and *Lipolexis,* having 2 spp. each, former parasitizing 2 spp. (2) of aphids on 2 spp.(2) of plants and latter attack 6 spp. (3) of host aphids, infesting 8 spp. (7) of plants. The rest of the parasitoids, *viz. Aeropraon Diaerietiella, Kashmiria, Lysaphidius, Lysiphelebia,* having one species each, parasitized 2 spp. of aphids (2 spp. of plants), 1 sp. (5 spp.), 1 sp. (1 sp.), 1 sp. (1 sp.) and 1 sp. (1 sp.) respectively. See Catalogue, p. 138 and Checklist, p.153.

The host aphid genus, *viz. Aphis,* with 1 Species, has yielded 8 different genera of aphidid parasitoids, infesting 31 spp. (24 genn.) of host plants. Next dominant aphid genera – *Chaitophorus* and *Macrosiphoniella,* have found to be parasitized by 6 and 5 genera of aphidiids on 6 and 7 Species of host plants respectively. *Chaetosiphon*, with 4 spp., are found to be parasitized by only genus *Aphidius* on 3 host plant species. The host aphids, *viz. Acyrthosiphon, Cavariella, Myzus* and *Uroleucon*, having 3 spp. each, yielded 3, 2, 3 and 5 different aphidiid genera on 10, 3, 7 and 4 plants species respectively. The host aphids such as *Brachycudus, Capitophorus, Cinara, Eriosoma Lachnus, Liosomephus, Melanaphis, Myzaphis Piriphyllus,* with 2 species each; yeilded 2, 2, 1, 3, 2, 4, 4, 4, 1 aphidiid genus/ genera, infesting 2, 2, 5, 3, 2, 2, 2, 2, 2 Species of host plant respectively. See catalogue, p. 138 and checklist, p. 153.

Rest of the aphid genera having one Species each, yielded highest number of parasitoid species as 5 in case of aphid genus *Macrosiphum,* followed by 4 pertaining to *Hyalopterus*, 3 spp. each with respect to *Eumyzus* and *Stagona* , infesting 5 spp. (3 genn.), 2 (2), 1(1) and 1 sp. (1. gen.) of host plant respectively. The aphid genera, other than mentioned above, have yielded either 2 or 1 aphidiid Species, infesting highest number as 3 spp. or 2 spp. or 1 sp. of host plants in different localities of Kashmir Valley (see Table 8.1, p. 130 and Checklist, p. 153).

TABLE 8.1. Species richness and diversity of aphid parasitoids (Hymenoptera) attacking host-aphid genera/species, infesting the number of host-plant species in Jammu and Kashmir

S. No.	Aphidiid Parasitoid Genera with No. of Species / Host Aphid Genera II	Adialytus (3 spp.)	Aphidius (16)	Aeropraon (1)	Binodoxys (2)	Diaeretiella (1)	Ephedrus (10)	Kashmiria (1)	Lipolexis (2)	Lysaphidius (1)	Lysiphlebia (1)	Lysiphlebus (3)	Monoctonus (3)	Pauesia (4)	Praon (7)	Toxares (4)	Trioxys (15)	Total No. of Host Species: Aphid	Total No. of Host Species: Plant
(a)	(b)	i	ii	iii	iv	v	vi	vii	viii	ix	x	xi	xii	xiii	xiv	xv	xvi	(c)	(d)
1	*Acyrthosiphon*	-	+	-	-	-	+	-	+	-	-	+	-	-	-	-	+	3	10
2	*Amphicercidus*	-	-	-	-	-	-	-	-	-	-	-	-	-	+	-		1	01
3	*Aphis*	-	+	-	-	+	+	-	+	-	-	+	-	-	+	+	+	9	31
4	*Aulacorthum*	-	-	-	-	-	+	-	-	-	-	-	-	-	-	-	-	1	02
5	*Betacallis*	-	-	-	-	-	-	-	-	-	-	-	-	-	-	+	-	1	01
6	*Brachycaudus*	-	+	-	-	-	-	-	-	-	-	-	-	-	-	-	+	2	02
7	*Brevicoryne*	-	-	-	-	+	-	-	-	-	-	-	-	-	-	-	-	1	02
8	*Capitophorus*	-	+	-	-	-	-	-	-	-	-	-	-	-	-	-	+	2	02
9	*Cavariella*	-	+	-	-	-	+	-	-	-	-	-	-	-	-	-	-	3	03
10	*Chaitophorus*	+	+	-	-	-	+	-	-	-	-	+	-	-	+	+	-	6	06
11	*Chaetosiphon*	-	+	-	-	-	-	-	-	-	-	-	-	-	-	-	-	4	03
12	*Cinara*	-	-	-	-	-	-	-	-	-	-	-	-	+	-	-	-	2	05
13	*Coloradoa*	-	+	-	-	-	-	-	-	-	-	-	-	-	-	-	-	1	01
14	*Eriosoma*	-	+	+	-	-	-	-	-	-	-	-	-	-	-	-	+	2	03
15	*Eumyzus*	-	+	-	-	-	+	-	-	-	-	-	-	-	-	+	-	1	01
16	*Hyalopterus*	-	-	-	-	-	+	-	-	-	-	-	-	-	+	+	+	1	02

Table contd...

(a)	(b)	i	ii	iii	iv	v	vi	vii	viii	ix	x	xi	xii	xiii	xiv	xv	xvi	(c)	(d)
17	*Hypermyzus*	-	-	-	-	-	-	-	-	-	-	-	-	-	+	-	-	1	01
18	*Impatientinum*	-	+	-	-	-	-	-	-	-	-	-	-	-	-	-	-	01	01
19	*Lachnus*	-	+	-	-	-	-	-	-	-	-	-	-	+	-	-	-	02	02
20	*Liosomaphis*	-	+	-	-	-	-	-	+	-	-	-	+	-	+	-	-	02	02
21	*Lipaphis*	-	-	-	-	+	-	-	-	-	-	-	-	-	-	-	-	01	01
22	*Macrosiphoniella*	-	+	-	-	-	+	-	-	-	-	-	+	-	+	-	+	06	07
23	*Macrosiphum*	+	+	-	+	-		-	-	-	-	-	-	-	+	-	+	01	06
24	*Melanaphis*	-	+	-	-	-	+	-	-	-	-	+	-	-	+	-	-	02	02
25	*Microlophium*	-	+	-	-	-	+	-	-	-	-	-	-	-	-	-	-	01	01
26	*Mindarus*	-		-	-	-	-	-	-	-	-	-	-	-	-	-	+	01	01
27	*Myzaphis*	-	+	-	-	-	-	-	-	-	-	-	+	-	+	-	+	02	02
28	*Myzus*	-	+	-	-	-	-	-	-	-	-	-	-	-	-	+	+	*03*	*07*
29	*Ovatus*	-	-	-	-	-	-	-	-	-	-	-	-	-	-	+	-	01	*01*
30	*Phorodon*	-	+	-	-	-	-	-	-	-	-	-	-	-	-	-	+	01	01
31	*Periphyllus*	-	+	-	-	-	-	-	-	-	-	-	-	-	-	-	-	02	02
32	*Prociphilus*	-	-	-	-	-	+	+	-	-	-	-	-	-	-	-	-	01	01
33	*Pterocomma*	-	+	-	-	-	-	-	-	-	-	-	-	-	-	-	-	01	01
34	*Rhopalosiphum*	-	+	-	-	-	-	-	-	-	-	-	-	-	+	-	-	01	01
35	*Schizoneurella*	-	-	+	-	-	-	-	-	-	-	-	-	-	-	-	-	01	01
36	*Semiaphis*	-	+	-	-	-	-	-	-	-	-	-	-	-	-	-	-	01	01
37	*Shivaphis*	-	-	-	-	-	-	-	-	-	-	-	-	-	-	-	+	01	01
38	*Sipha*	-	-	-	-	-	-	-	-	-	-	+	-	-	-	-	-	01	03
39	*Sitobion*	-	+	-	-	-	-	-	-	+	+	-	-	-	-	-	-	02	01

Table contd...

(a)	(b)	i	ii	iii	iv	v	vi	vii	viii	ix	x	xi	xii	xiii	xiv	xv	xvi	(c)	(d)
40	*Stagona*	-	-	-	-	-	-	-	-	-	-	-	-	-	+	-	-	01	01
41	*Tetraneura*	-	-	-	-	-	+	-	-	-	-	-	-	-	-	-	-	01	01
42	*Therioaphis*	-	-	-	-	-	-	-	-	-	-	-	-	-	-	-	+	01	03
43	*Tuberolachnus*	-	+	-	-	-	-	-	-	-	-	-	-	-	-	-	-	01	01
44	*Uroleucon*	-	-	-	+	+	+	-	-	-	-	-	-	-	+	+	-	03	04
45	Unidentified . Aphidgen	+	-	-	-	-	-	-	-	-	-	-	-	-	-	-	-	02	02

.+ denotes, parasitization of aphid genera/species by aphidiid parasitoid genera/species

- indicates no parasitization by aphidiids.

8.1.2.3 Braconids, chalcids and cynipids (faunistic survey, diversitry and as biocontrol agents)

I. ***Braconids:*** Braconids (Braconidae: Ichneumonoidea) are relatively small-sized and stout-bodied wasps, with abdomen about as long as the head and thorax combined. The adults visit flowers and are often attracted to light. In majority of braconids, pupation takes place outside of the body of the host or spin cocoons away from the host. In *Apanteles* spp., pupation takes place externally and the pupa remains enclosed in a cocoon, formed of fine threads of white, yellow or cream coloured silk.

From various regions of J & K State, so far, a total of 14 Species of Braconids has been reported, distributed over 6 genera. These included 8 spp. belonging to genus *Apanteles*, 2 spp. to *Phaenocaroa* and 1 sp. each to *Blacus, Bracon, Protatus* and *Spathius*. See Catalogue, pp. 149-150.

Previous studies on braconid parasites in Kashmir Valley, attacking gypsy moth (*L. obfuscata)* and maize borer (*O. kashmirica),* for biological control of these forestry and agricultural pests in U.S.A, have been carried out by the entomologists of Commonwealth Institute of Biological Control (CIBC) during the year 1968 and 1969. It was reported that a total of 3 braconids, *viz. Apanteles conspersae, A. liparidis* and *Apanteles* sp., suppressed the population of *L. obfuscata* and, *Bracon brevicornis* and *Apantales* as larval parasites of *O. kashmirica* prevalent from field samples, collected from localities like Larikpora and Achabal (Anon., 1969, 70).

During the course of extensive surveys for the natural enemies of *L.obfuscata*, Zutshi (1967) and Rishi & Shah (1985), recorded more braconid parasites, belonging to the genus *Apanteles, viz., A. flavicoxis, A. indiansis, A. melanoscelus, A. porthetriae* and *A.* nr. *solitarius*. See Catalogue, p.149.

Two species including one new species, pertaining to *Phaenocarpa,* under sub-family Allysiinae, tribe Allysini, have been recorded by Bhat (1979) from Gulmarg, Pahalgam, Khilanmarg and Sonamarg localities. Sharma & Gupta (1985) described a new doryctine genus, *Protatus,* parasitizing *Sinoxyla* sp., a serious pest of *Dalbergia sissoo* in Jammu region. A new Species and first generic record of a braconid from India, *viz. Blacus* (B.) *imitator*, has been given by Papp (1985). See Catalogue, (pp. 149-150).

Pawar & Parry (1989) recorded the occurrence of *Apanteles* sp., suppressing the population of *Archips subsidiaria,* a major pest of apple in Nimmu region in district Leh of Ladakh. Recently, Buhroo *et al.* (2001) found *Spathius* sp., parasitizing larvae/pupae of shot-hole borer, *Scotylus nitidus* in fruit-growing area of Kashmir Valley.

II. ***Chalcidids:*** Chalcidids have been placed in the family Chalcididae under Super-family Chalcidoidea. Theses are primary parasites of Lepidoptera, Diptera, Hymenoptera Coleoptera besides, bugs are also attacked. Some are also hyperparasites, attacking tachinids and ichneumonids.

Earliest report of chalcidid, *viz. Brachymeria euloeae*, attacking *L. obfuscata* on poplars from Kashmir, has been given by Zutshi (1967). Later, chalcidid, *B. intermedia*, as pupal parasite of *L. obfuscata* on salix and poplar from Srinagar, has been reported by CIBC scientists, during the course of their project for evaluation of hymenopteran parasites for biological control of pests in U.S.A. (Anon., 1969, 70).

In the recent years, Rishi & Shah (1985) and Masoodi *et al.* (1986), reported the existence of *B. lasus*, parasitizing pupae of *L. obfuscata* on salix and poplars in Kashmir. In addition to the above mentioned three Species of genus *Brachymeria*, 3 more Species, have been reported from Kashmir region, these included *Brachymeria tachardia, Chalcis ornatipes* and *Epitrans kashmirensis* by Subba- Rao & Hyat (1986) Cameron (1906 and Husain & Agarwal (1981) respectively. See Catalogue, p. 141.

III. ***Cynipids:*** Cynipids (Cynipidae) are included in the Super-family Cynipoidea, commonly called as gall-wasps, because majority of the Species produce galls in plants (sub-family cynipinae). However, many Species belonging to sub-family Eucolinae and Charipinae are parasitic. Charipinae includes Species which are mainly hyperparasites of aphids through primary parasites (aphidiidae) or other parasites. The cynipids are minute insects, black or brown, with abdomen globose or compressed and shiny.

In Kashmir Valley, two Species of cynipids, *viz. Alloxysta* sp. and *Phaenoglyphis*, have been recognized as hyperparasites, attacking aphidiids of different aphid hosts. Bhagat (1983, 87), gave records and host range (parasite, aphid and plants), from different ecological habitats of these two Species, with special emphasis on occurrence of hyperparasites of primary parasities of aphids infesting *Rosa* spp.

Alloxysta spp. have been found in various localities *ca* 1500 m to 2,500 m, parasitizing the primary parasite, *viz., Praon volucre, Aphidius eglanteriae* on aphid, *Myzus rosarum,* attacking mainly host plant, *Rosa, webbiane* (wild rose). Besides, it also attacked *Aphidius* spp. (4), *Lysiphlebus ambiguous, Toxares* spp. (2) on 8 spp. each of host aphid and plant. *Phaenoglyphis* sp. was observed as hyperparasite of aphids *Microlophium carnosum* and *Aphis* spp, attacked by primary aphidiid parasitoids, *Aphidius* spp. (2) and *Praon abjectum* on host plants, *Urtica dioica* and *Nepeta salvifolia.* See Catatogue, p.144.

8.1.2.4 Encyrtids, Eulophids and Eupelmids

I. ***Encyrtids:*** Encyrtids (Encyrtidae: Chalcidoidae) are endoparasites as well as hyperparasites of aphids, scale insects and whiteflies. They have robust body, showing broad convex mesopleura. From Kashmir, encyrtid, *viz. Microterys chalcostomus,* has been reported as a endoparasite of scale insect, *Lecanium. Aphidencyrtus aphidivorus* and *Prionomitus* sp. are observed to be as hyperparasites of aphids in this region. See catalogue, p.141.

Bhagat (1982, 83, 87) made thorough field studies on host records and range of the encyrtids (*A. aphidivorus* and *Prionomitus),* with reference to gall-forming aphids and aphids infesting *Rosa* spp. *A. aphidivorus* has been found to parasitize 3 spp. of primary aphidiid parasitoid, *viz. Aphidius* sp. *Praon* sp. and *Protaphelinus nickolskajae* on different host aphids, infesting 3 host plant Species. In case of *Prionomitus,* primary aphidiid parasitoid Species included *Aeropraon lepelleyi, Aphidius* spp. and *Trioxys indicus* on four different aphid and host plant Species. See Catalogue, p.138 and Host-checklist, pp.161-162.

II. ***Eulophids:*** Eulophids belong to the family Eulophidae under Superfamily Chalcidoidae, parasitic wasps, small to very small size (16 mm). Some are as hyperparasite and a few phytophagous Species. The eulophids attack scale insects, concealed Lepidopterans and Dipterans (leaf-miners). From Srinagar, one Species of Eulophid, *viz. Chrysocharis nitetis* has been reported by Hansson (1985). See Catalogue, p.141.

III. ***Eupelmids:*** Eupelmids have been placed under the family Eupelmidae and Superfamily Chalcidoidae. The female body is elongate, frequently metallic coloured and often 'u' shaped in dried specimen. The larvae of eupelmids feed ectoparasitically and also endoparastically, showing wide host spectrum.

The genus *Anastatus,* having three Species (*A. bifasciatus, A. kashmirensis* and *Anastatus* sp.), reported from Kashmir. These species have been bred from eggs of Lepidopteran pests: *L. obfuscata* on *Salix* spp. and *Populus* spp. and *Porthetria dispar* infesting *Salix* spp. *and Populus* spp. Apart from these three Species from Kashmir, 3 more Species belonging to the genus *Eupelmus (E. kashmirensis, E.valsus* and *E. vindox)* have been found to parasitize the larvae of Coleopteran, *Scolytus* spp. (shot hole borer) on apple trees. (see Catalogue, pp. 141-142).

Earlier, the programme for the evaluation of hymenopteran parasites of *L. obfuscate* in Kashmir region by CIBC scientists for bio-control of pest in U.S.A, revealed the prevalence of *Anastatus* spp. (*A. kashmirensis* and *A. bifasciatus)* from the eggs of *L. obfuscata* as early as sixties (Rao, 1966 and Anon, 1969). Later surveys by Rishi & Shah (1985) and Masoodi *et al.* (1986) confirmed the wide prevalence of these effective egg parasites of gypsy moth from different parts of the valley. Recently, Narendran *et al.* (2001) reported the existence of three Species of *Eupelmus,* as effective parasite of shot-hole borer beetles, damaging apple trees in many orchards of Kashmir region.

8.1.2.5 Eurytomids, Ichneumonids and Megaspilids

I. ***Eurytomids:*** Eurytomids belong to the family Eurytomidae under Super-family Chalcidoidea, commonly known as seed chalcids usually black in colour, with thorax, head and antennae often hairy. From Kashmir, two Species of eurytomids, *viz. Eurytoma morio* and *Eurytoma* sp., as larval endoparasites of *Scolytus* sp. on apple tree and *L. obfuscata* on *Salix* spp., have been reported by Narendran *et al.* 2001 and Masoodi *et al.* (1906) respectively (see catalogue, p. 142).

II. ***Ichneumonids:*** The ichneumonids are mostly slender wasps, belonging to the family Ichneumonidae and superfamily Ichneunoidae, and is the largest family of parasitic hymenoptera, characterized by different size, form, colouration, presence of long antenna and in many ichneumonids, the ovipositor is long, often longer than the body. All Species are parasitic on larval and pupal holometabolous insects. The parasite may complete its development in the stage of host in which the egg is laid or in later stage. Most Species spin cocoon and pupate, where the host died or pupate within the cocoon or pupae of host.

In this region, about 38 Species of ichneumonids, belonging to 26 genera are known to occur, parasitizing larval, prepupal stages of gypsy moth *(Lymantria obfuscata)* (serious forest pest) wood-boring coleopterans, sawflies, agricultural pests *Plutella xylostella (= maculipennnis)* and *Ostrinia kashmirica,* besides other unknown insect Species pertaining to orders like Lepidoptera, Hymenoptera and Coleoptera. See catalogue, pp.150-152.

Of the various genera of ichneumonids reported from J & K, the dominant genus having 5 Species is *Scambus,* followed *Itoplectis, Paniscus* and *Pimpla,* with 3 Species each, the next dominant genera, having 2 Species each are *Theronia* and *Xorides*. The rest of the known genera of this region, are having 1 Species each. See catalogue, pp.150-152.

Lymantria obfuscata is responsible for causing much damage to salix and poplars besides, fruit trees in Kashmir, has been found to be suppressed by 6 Species of ichneumonds, *viz. Anilastus* sp., *Coccygomimus* sp., *Itoplectis himalayensis, Pimple* sp. *Theronia atalantae atalantae* and *Theronia* sp. First report of ichneamonid, *viz. Pimpla* sp., parasitizing the pupae of *L. obfuscata* was given by Zutshi (1967). See catalogue, pp.150-151.

During the year 1968 and 1969, the evaluation of ichneumonid parasites as biological control agents of *L. obfuscata* was carried out by the scientists of Commonwealth Institute of Biological Control (CIBC) London. Various parasites recorded and reared in Kashmir (for U.S.A.) included *Anilastus* sp., *Coccygominus* and *Theronia atalantae atalantae.* (Anon., 1970). During the recent years, Rishi & Shah (1985), and Masoodi *et al.* (1986), have made detailed field observations on the ichneumonid parasites of gypsy moth in Kashmir.

In Kashmir, the scientists of CIBC, during 1968-69 made thorough field studies on the parasites of major insect pest of agricultural importance, *Ostrinia kashmirica* (Pyralidae: Lepidoptera), for biological control of *Ostrinia* major pest of agricultural importance in U.S.A. The samples were collected from localities, *viz.* Larkipora and Achabal. The larval parasites of *O. kashmirica,* reported from the field samples included *Exeristes roboraton, Pristomerus* and *Scambus flaricrus* (Anon, 1970).

Simmonds & Rao (1960), for the first time recorded two ichneumonids, *viz. Horogenes* sp. and *Itoplectis* sp. from Srianagar, as parasites of major agricultural pest, *Plutella xylostella* .

Various workers who made significant contribution to taxonomic studies of ichneumonid-fauna of J & K are: Gupta (1967, 78, 80, 85 & 91); Gupta & Chandra (1974), Gupta & Tikar (1967); Diller (1984) and Horstmann (1992). These authors not only gave new records (2 spp.) but also described 13 new Species, belonging to different genera, *viz. Deuteroxorides* (1 sp.), *Ecthrus* (1 sp.) *Eridolius* (1 sp.), *Inocellia* (1 sp.) *Itoplectis* (1 sp.), *Poemenia* (1 sp.), *Polyblastus* (1 sp.), *Promethes* (1 sp.), *Scambus* (3 spp.) and *Xorides* (2 spp.), parasitic on pupae of Lepidopterans, larvae/pupae of hard wood/coniferous tree boring coleopterans and larvae of sawflies (Tenthredinidae) as ectoparasitoids, distributed in regions like Batalik (Ladakh), Batote (Jammu), Gulmarg, Pahalgam, Shopian and Srinagar (Kashmir) (For delails, see Catalogue, pp. 150-152 and Host Checklist, p.153).

III. ***Megaspilid:*** Megaspilids belong to the family Megaspilidae and superfamily Ceraphronoidea, are small insects and mostly hyperparasites or secondary parasites of aphids through primary parasites as aphidiids. In Kashmir Valley, Bhagat (1983) reported the existence of Megaspilid, *viz. Dendrocerus carpenteri* and studied its host range. *D. carpenteri* was found to develop as hyperparasites of aphidiid *viz. Aphidius* spp. on aphids (*Macrosiphum rosae* and *Uroleucon sonchi)* and *Praon* spp. on aphids *(Aphis grossulariae* and *Amphicercidus tuberculatus),* infesting four Species of host plants *(Rosa webbiana, Sonchus* sp., *Lonicera quinquelocularis* and *Epilobium hirsutum),* See (Catalogue, p.139 and Host Checklist, p.153).

8.1.2.6 Mymarids, Pteromalids, Scelionid, Torymids and Trichogrammatids

I. ***Mymarids:*** Mymarids (Mymaridae: Chalcidoidea) are commonly known as fairflies, extremely minute in size (0.2 mm), mostly black or yellow, lacking metallic colour,These are exclusively egg-parasited of insects, preferring host eggs in concealed habitat, such as plant tissues under bark and in soil. Some Species have been found as successful agents 'n biological control programme.

In J & K State, five Species of mymarids, belonging to genus *Gonatocerus* (4 spp.) and *Paralletaptera* (1 sp.), have so far been reported by Zeya & Hayat (1995) and Subba Rao & Hayat (1986) respectively. These Species are known to be as egg-parasites of various unknown insect Species on unknown host plant Species, from various regions like Jammu and Kashmir (Pahalgam and Anantnag), see catalogue, p. 142.

II. *Pteromalids:* Pteromalids are minute-sized, black or metallic green or bronze coloured insects, belonging to the family Pteromalidae under super-family Chalcidoidea. These are important entomophagous form of species, attacking wide variety of insects of different orders. Most of these are primary parasites and many are valuable in the biological control of crop pests. Many Species are also secondary/hyperparasites. The most common Species in Kashmir are belonging to genus: *Asaphes* spp., *Coruna* spp. and *Pachyneuron* spp. The female of hyperparasite Species attack/parasitized aphid that include last instar larva or pupa of the aphidiid (primary parasitoid).

In Kashmir, about 11 Species of pteromalids, spread over 8 genera, have been reported as primary parasites and hyperparasites of insect hosts– *Scotytus* spp. on apple fruit tree and aphids on wide range of host plant Species respectively. Various pteromalid genera, with number of Species include *Asaphes* (2 spp.), *Coruna* (2 spp.)., *Cheiropachus* (1 sp.), *Heydenia* (1 sp.), *Macromesus* (1 sp.), *Mesopolobus* (1 sp.), *Pachyneuron* (2 spp.) and *Rhaphitelus* (1 sp.). The Species of pteromalid parasites are listed in the catalogues, with their host insects (primary/secondary) and host plant Species along with information on authority reporting the Species, see Catalogue, p.142-143 and Host Checklist, p.153.

Bhagat (1983, 87) recorded and studied the host range of pteromalids as hyperparasitoids of aphids on wide range of host plants, these include *Asaphes suspensus* (Nees) *Coruna* sp., *Pachyneuron aphidis* and *P. groenlandicum* (Holmgren). *A. suspenscus* was found to be hyperparasitoid of 7 spp. primary parasitoids (*Aphidius* spp. 4, *Diaertiella rapae, Proan volucre* and *Trioxys* sp.) on 8 Species of aphids, infesting 8 host-plant Species. *Coruna* sp. attacked 5 Species of primary aphidiid parasitoids of 6 Species of aphids on host-plants, *viz. Rosa* spp. (3), *Aesculus indicus, Lonicera quinquelocularis* and *Prunus domestica insititia.* In case of *Pachyneuron* spp. *(P. aphidis, P. groenlandicum)*, each having 7 Species of host aphidiid parastoids and aphids, on 6 host plant Species. Apart from this, Boucek *et al.* (1978) reviewed the pteromalidae of this region and reported *Coruna clavata* Walker as hyperparasitoid of unknown aphidiid parasitoid on aphid, *viz. Macrosiphum* sp., see Host Checklist, p.153.

Recently, Buhroo *et al.* (2001) and Narendran *et al.* (2001), have recorded for the first time two Species of pteromalids (*Cheiropachus quadrum, Rhaphitelus maculata)* and also described two new Species– *Heydenia indica* and *Macromesus harithus*, as economically important parasite of shot-hole borer, *Scolytus nitidus* (Scolytidae: Coleoptera), damaging apple trees and fruit trees in many orchards of Kashmir valley.

III. ***Scelionids:*** Scelionid wasps (hymenoptera) belong to family Scelionidae and Superfamily Proctotrupoidea. Adult is non-metallic, sclerotized, solitary, internal egg parasite of hemiptera, prefers eggs in aggregation clumps. In this region, Scelionid,*viz. Hadronotus,* has been recorded as a parasite of egg of *Cletus bipunctatus* (Coreidae: Hemiptera) (Mathur & Sharma,1962).

IV. ***Torymid:*** Torymid wasps belong to the family Torymidae under superfamily Chalcidoidea. The body of Torymid wasp is moderate seized and typically metallie blue or green.These wasps are ectoparasite of hymenoptera and coleoptera, some as phytophagous. In Kashmir valley, torymid, *Monodontomerus acreus* has been found suppressing the population of *L. obfuscata*, a major pest of forestry importance and was reared in Srinagar by CIBC scientists during 1968, for use as bio-control agent against gypsy moth in U.S.A. (Anon 1970), see catalogue, p. 143.

V. ***Trichogrammatids:*** Trichogrammatid wasps have been placed under family Trichogrammatidae and super-family Chalcidoidea, are smallest insects (0.2–1.0 mm), weakly sclerotized, without metallic colouration, showing 3-segmented tarsi and presence of microscopic rows of hairs on the wings. These are parasites of eggs of insects and genus *Trichogramma*, usually parasitize Lepidopterans. Trichogrammatids are being extensively utilized in biological control programmes in different parts of the world.

Pawar *et al*. (1981) conducted field study in the apple orchards in Ladakh region to determine the effectiveness of two Species of Trichogrammatid (exotic) pertaining to the genus *Trichogramma, viz. T. cacoeciae pallidium* and *T. embryophagum,* against codling moth *(Cydia pomonella)*, damaging apple. With the introduction of these two exotic egg parasites of *C. pomonella*, reduction in the incidence of codling moth has been achieved from 17.5–37.5%. The scientists have confirmed the effectiveness of the egg-parasitoids in Ladakh region against insect pest infestations and they further suggested that it would be useful if weekly releases of these bio-control agents were made against the moth (pest) from the month of June to July.

8.2. ENTOMOPHAGOUS PARASITIC INSECTS, HOST INSECTS AND PLANT SPECIES CATALOGUE

[Keys to the abbreviations and Roman numerals, denote author(s)/reference(s), given at the end of the Catalogue].

(I) *Dipteran Parasites/Parasitoids*

(A) Superfamily: Conopoidea

(a) Family: Conopidea

(1) *Ceria dimidatipennis* Brunetti (XXIVA, Kmr.) Adult, bees, wasps, orthoptera

(A) Superfamily: Muscoidae

(a) Family: Muscidae

(1) *Muslina stabulanus* (Fab.) (LXII, Kmr.).

Host: *Lymantria obfuscata* on *Salix* spp. and Poplars.

(b) Family: Sarcophagidae

(1) *Sarcophaga* sp. (LXXVII, LXII, Kmr.)

Host: *L. obfuscata* on *Salix* sp.

(c) Family: Tachinidae

(1) *Corcelia* sp. (III, Srg.)

Host: *L. obfuscata* on *Salix* and Poplars.

(2) *Compsilura* sp. (XXXXV, LX, Kmr.)

Host: *L. obfuscate* on *Salix* spp.

(3) *Drino discreta* (III, XXXXV, LX, Srg.)

Host: *L. obfuscata* on *Salix* spp. & *Populus* spp.

(4) *Drino incompicuoides* (LXXVIII, Kmr).

Host: *L. obfuscata* on *Salix.*

(5) *Exorista rossica Mesnil* (III, LXII, LXXVIII, Kmr.)

Host: *L. obfuscata* on *Salix* spp.

(6) *Voria ruralis* spp. (LXX, Srg.)

Host: *Plutella xylostella (= maculipennis)* on unknown host plant species).

(II) *Hymenopteran Parasites/Parasitoids*

(A) Superfamily: Ceraphronoidea

(a) Family: Megaspilidae

(1) *Dendrocerus carpenteri* (Curtis) (XII, Kmr.)

Host: *Aphidiid parasitoids; Aphidus avenae–Macrosipham rosae* on *R. webbeana; Aphidius* sp.*–Uroleucon sonchi* on *Sonchus* sp., *Praon dorsale–Amphicercidus tuberculatus* on *Lonicera quinquelocularis; Praon–objectum–Aphis grossulariae nr. pollinosa* on *Epilobium hirsutum.*

(B) Superfamily: Chalcidoidea

(a) Family: Aphelinidae

(1) *Aphelinus (Aphelinus) abdominalis* (Dalman) (XXXIV, LXXIV, J & K)

Host: *Aphis craccivora* on unknown host plant.

(2) *Aphelinus (Aphelinus) asychis* Walker (XXXIV, LIX, LLXXIV, Kmr.)

Host: *Brevicorynae brassicae, Lipaphis erysimi, Myzus persicae* on unidentified host plants.

(3) *Aphelinus (Aphelinus) flavipes* (Forester) (XXXIV, LIX, LXXIV, J & K).

Host: *Aphis gossypii, A. spiraecola, L. erysimi, M. persicae,Rhopalosiphum* sp.

(4) *Aphelinus* nr. *fulvus* Jashnosh (XIV, XV, Duc., Khi., Kmr.)

Host: *Drepanosiphum plantantoids* (Schrauk) on *Acer caesium; M. cerasi* on *Spirea* sp.

(5) *Aphelinus kashmiricus* Hayat (XIV, XXXIX, Srg.)

Host: *Aphis gossypii* on *Zinnia elegans*

(6) *Aphelinus mali* Haldman (VIII, XIII, XIV, LI, LXXVI, Kmr.)

Host: *Aphis affinis* on *Epilobium hirsutum; A. craccivora* on *R. pseudoacacia;*

A. kurosawai on *Dalphne oleoides; A. nerii* on *Verbascum thapsus; E. lanigerum* on *M. pumila; Sitobion avenae (=Macrosiphum avenae)* on *Pennisetum flaccidum.*

(7) *Aphelinus* sp. (XXII, Kmr.)

Host: *Eulecanium coryli* on *Prunus armeniaca, Pulvinaria inconspiqna* on *C. australis*

(8) *Aphylis diaspidis* How. (Rao *et al.*, 1971) (Kmr.)

Host: *Quadraspidiotus perniciosus* on deciduous fruit trees, *Prunus* spp. and *Salix* spp.

(9) *Aphytis lingnanensis* Compere (XXII, Kmr., Srg., Pat., Han.)

Host: *Parthenolecanium* Corni (Bouche) on *Viburnum foetans, C. perniciosus on M. pumila.*

(10) *Aphytis paramaculicornis* DeBach & Rosen (XXII, XXXXIV, Han., Pat., Srg., Ana.)

Host: *Lepidosaphes kirgisica* on *Salix fragilis, Q. perniciosus* on *M. pumila*

(11) *Aphytis* sp. *proclia* group (XXXXVI, Kmr.)

Host: *Q. perniciosus* on *M. pumila.*

(12) *Azotus kashmirensis* Narayanan (LII, LXXIV, Kmr.)

Host: *Q. perniciosus.*

(13) *Coccophagus tschirchii* Mahdhihassan (LXXIV, Kmr.)

Host: *Ceroplastodes* sp. on undertermined plant species.

(14) *Encarsia kashmirensis* Singh *et al.* (LXXII, Sha.)

(= *Prospaltella perniciosi* Tower)

Host: Egg of *L. obfuscata* on fruit tress (*Prunus* spp.)

(15) *Encarsia perniciosi* (Tower) (Rao *et al.*, XXII, XXXXVIII, Kmr.) (= *Prospaltella perniciosi* Tower).

Host: *Q. perniciosus* on deciduous fruit trees, *Prunus* spp., *M. pumila* and *Salix* spp.

(16) *Marietta carnesi* (Howard) (LII, LXXIV, J&K)

(= *Marietta indica*)

Host: *Q. perniciosus*

(17) *Protaphelinus dargahae* Rishi & Lone (LXI, LXXIV, Srg.)

Host: *Pemiphigus bursarius* on *Populus nigra* var. *italica.*

(18) *Protapheinus nickolskajae* Joshnosh (Mackauer, 1972, VIII, XIV, XXXIX, IIX, LXXIV, Kmr.)

Host: *Epipemphigus imaicus* on *Populus ciliata, Pemphigus aedificator* on *P. ciliata, P. immunis* on *Populus caspica, P. ciliata* and *P. nigra, P. mordvilkoi* on *P. caspica, P. napeus* on *P. alba.*

(19) *Pteroptrix chinensis* (Howard) (LII, LXXIV, J & K)

Host: *Q. perniciosus.*

(b) Family: Chalcididae

(1) *Brachymeria euloeae* Wester (LXXVIII, Kmr.)
Host: *L. obfuscata* on *Salix* sp.

(2) *Brachymeria intermedia* (Nees) (II,III, LXII, LXXIV, Kmr.)
(= *Chalcis intermedia* Nees)
Host: *L. obfuscata* on *Salix* spp. and *Populus* spp.

(3) *Brachymeria lasus* Walker (XXXXV, LXII, Kmr.)
Host: *L. obfuscata* on *Salix* spp.

(4) *Brachymeria tachardia* (Cameron) (LXXIV, Kmr.)
Host: Unknown insect and plant species.

(5) *Chalcis ornatipes* Cameron (XXVI, Kmr.)
Host: Unknown plant host and insect species.

(6) *Epitranus kashmirensis* Hussain & Aggarwal (XXXXII, Srg.)
Host: Unknown host insect and plant species.

(c) Family: Encyrtidae

(1) *Aphidencyrtus aphidivorus* (Mayr) (VIII, XII, Kmr.)
Host: Aphid parasitoids: *Aphidius* sp. – *Ovatus* nr. *crataegarius* on *Mentha longifolia; Praon* sp. – *Chaetosiphon* sp. on *Rosa webbiana; Protaphelinus nickolskajae– Pemphigus* on *P. ciliata.*

(2) *Microterys chalcostomus* (Jalman) (LXXIV, Kmr.)
Host: *Lecanium* on undetermined host-plant species.

(3) *Prionomitus* sp. (XII, XIX, Kmr.)
Host: Aphidiid parasitoids: *Aeropraon lepelleyi* on aphid, *viz. Schizoneurella indica* on *Ulmus wallichiana; Aphidius* spp. – on *Aphis farinosa* on Salix alba, *Chaetosiphon tetrarhodus* on *Rosa canina* and *Rosa* sp.; *Trioxys indicus* on *Aphis affnis* on *M. longifolia (= M. sylvestris).*

(d) Family: Eulophidae

(1) *Chrysocharis nitetis* (XXXVIII, Srg.)
Host: Unknown host insect and plant species.

(e) Family: Eupelmidae

(1) *Anastatus bifasicatus* (XXXXV, LX, Kmr).
Host: Eggs of *L. obfuscata* on *Salix* spp. *Populus* spp.

(2) *Anastatus kashmirensis* Mathur (XXXX, XXXXV,IL, LXII, LXXIV
Host: *L. obfuscata* on *Salix* and *Populus*, *P. dispar.*

(3) *Anastatus* sp. (II, LX, LXII, LXXI, LXXVIII, Kmr).
Host: Eggs of *L. obfuscate* on *Salix* spp. and *Populus* spp., *P. dispra.*

(4) *Eupelmus kashmiricus* Narendran *et al.* (IV, Kmr.)
Host: *Scolytus* sp. on *M. pumila (= Malus sylvestris)*

(5) *Eupelmus valsus* Narendran *et al.* (IV, Kmr.)
Host: *Scotytus* sp. on *M. pumila.*

(6) *Eupelmus vindox* Erdos (XXV, IV, Kmr.)
Host: *Scotylus* sp. on *M. pumila.*

(f) Family: Eurytomidae

(1) *Eurytoma morio* Boheman (XXV, LV, Kmr.)
Host: *Scolytus* sp. on *Malus pumila (=M. sylvestris)*

(2) *Eurytoma* sp. (XXXXV, Kmr.)
Host: *L. obfuscata* on *Salix* sp.

(g) Family: Mymaridae

(1) *Gonatocerus longicornis* Nees (LXXVII, Jmu.)
Host: Egg of unknown insect species on unknown host plant.

(2) *Gonatocerus munnarus* Mani & Saraswat (LXXVII, J&K)
Host: Egg of insect species on unknown host-plant.

(3) *Gonatocerus pahlgamensis* (Narayanan) (LXXVII, Pah.)
(= *Lymaenon pahlgamensis* Narayanan)
Host: Egg of unknown insect species on unknown plant.

(4) *Gonatocerus tarae* (Narayanan & Subba Rao) (LXXVII, J & K)
Host: Egg of insect on unknown host-plant species.

(5) *Paralletaptera anantanagona* (Narayanan) (LXXIV, Ana.) (= *Polynema anatanagona* Narayanan)
Host: Unknown.

(h) Family: Pteromalidae

(1) *Asaphes fletacheri* (Crawford) (LXXVI, Kmr.)
Host: *Eriosoma lanigerum* (aphid) on *Malus pumila.*

(2) *Asaphes suspensus* (Nees) (XII, XIX, LXXIV, Kmr.)
Host: Aphidiid parasitoids: *Aphidius areolatus – Periphyllus aesculi* on *Aesculus indicus; A. avenae – Chaetosiphan chaetosiphon* on *R. macrophylla,* on *Rosa indica; A. matricariae – Capitophorus elaeagni* on *Carduus edelbergi; A. rosae – C. chaetosiphon* on *R macrophylla; Dieretiella rapae – Breviconyne brassicae* on *Brassica oleracea; Proan volucre – Macrosiphum* sp. on *Rosa* sp; *Trioxys indicus – Aphis fabae solanella* on *Solanum nigrum (= miniatum), Aphis affinis* group on *M. longifolia.*

(3) *Coruna clavata* Walke (XXIV, LXXIV, Kmr.)

Host: Aphidiid parasitoids on aphid *(Macrosiphum)*

(4) *Coruna* sp. (VIII, XII, XVII, XIX, Kmr.)

Hosts: Aphid parasitoids: *A. areolatus – P. aesculi* on *A. inidcus; Aphidius* spp. – *Chaetosiphon* sp. on *Rosa* sp., *M. rosae* on *Rosa* sp., *Macrosiphun* sp. on *Rosa indica; Ephedrus srinagarensis – Prociphilus* sp. on *L. quinquelocularis; P. volcure – Hyalopterus pruni* on *Prunus domestica insititia; Praon* sp. – *Chaetosiphon* sp. on *R. macrophylla.*

(5) *Cheiropachus quadrum* (Fabricus) (XXV, LX, Kmr.)

Host: *Scolytus* sp. on *M. pumila.*

(6) *Heydenia indica* Narendran *et al.* (XXV, LV, Kmr.)

Host: *Scotytus* sp. on *M. pumila.*

(7) *Macromesus harithus* Narendran *et al.* (LV, Kmr.)

Host: *Scotytus* sp. on *M. pumila.*

(8) *Mesopolobus beesoni* (Mani & Kaul) (XXXXIII, LXXIV, Kmr.)

Host: unknown host and plant insect species.

(9) *Pachyneuron aphidis* (Bouche) (XII, XVII, XXIV, LXXIV, Kmr.)

Hosts: Aphid parasitoids: *Aphidius gifuensis – A. fabae* on *Capsella bursa pestoris; A. matricariae – C. edelbergi* on *C. edelbergi; A. uzbekistanicus – Macrosiphum avenae* on *Sorghum halepense; Aphidius* sp. – *Aphis farinosa* on *Salix aegyptica; P. volucre – H. pruni* on *P.d. insititia; Trioxys acalephae – M. rosae* on *R. webbiana; Trioxys rishii – C. elaeagni* on *C. edelbergi.*

(10) *Pachyneuron groenlandicum* (Holmgren) (XII, XXIV, LXXIV, Kmr.) (= *Pachyneuron karnalensis* Mani)

Hosts: Aphid parasitoid: *Trioxys rishii – C. elaeagni* on *C. edellergi; Brachycaudus cardui* (aphid).

(11) *Rhaphitelus maculats* Walker (XXV, LV, Kmr.)

Host: *Scolytus* sp. on *M. pumila.*

(i) Family: Torymidae

(1) *Monodontomerus acreus* (III, Kmr.)

Host: *L. obfuscata* on unknown host-plant.

(j) Family: Trichogrammatidae

(1) *Trichogramma cacoeciae pallidium* (Mayer) (LVII, Ldk.)

Host: *Cydia pomonella* on *M. pumila* (apple cultivated).

(2) *Trichogramma embryophagum* (Hartig) (LVIII, LXXIV, Ldk.) (= *Encyrtus embryophagum* Hartig).

Host: *C. pomenella* on *M. pumila.*

(C) Superfamily: Cynipoidea

(i) Family: Cynipidae

(1) *Alloxysta* sp. (XII, XIX, Kmr.)

Hosts: Aphidiid parasitoid: *Aphidius uzbekistanicus – Impatientinum impatiens dalhousiensis* on *Impatiens, (Macrosiphum avenae* sp., *Sitobion avenae (=Macrosiphum avenae)* on *Rosa* sp., *Macrosiphum* sp. on *Rosa brunoni; Lysiphlelus ambiguus – Sipha maydis* on *Poa angustifolia; P. volucre – Myzaphis rosarum* on *Rosa* sp., *Toxares deltiger – Betacallis* sp. on *Conium maculatum; Toxares zakai – Microlophium* sp. on *urtica dioica.*

(2) *Phaenoglyphis* sp. (XII, Kmr.)

Hosts: Aphidiid parasitoid: *Aphidius ervi – Microlophium carnosum* on *U. dioica; Aphiidius gifuensis – Aphis* sp. on *Nepeta salvifolia.*

(D) Superfamily: Ichneumonoidea

(a) Family: Aphidiidae

(1) *Adialytus ambiguus* Halday (XXI, Bat., Raj.)

Host: *Macrosiphum* on *Songhum* sp.

(2) *Adialytus arvicola* (Stary) (LXVII, Sop.)

Host: Aphid on unknown host-plant.

(3) *Adialytus salicaphis* (Fitch) (LVII, LXVII, Sop., Kmr., Srg.)

Host: Aphid on unknown host plant species; *Chaitophorus kapuri* on *Populus* sp.

(4) *Aphidius absinthi Marshall* (XXI, Jmu.)

Host: *Macrosiphoniella* sp. on *Chrysanthemum* sp.

(5) *Aphidius areolatus* Ashmead (XII, XV, Kmr.)

Host: *P. aesculi* on *A. indicus; P. vandenboschii* on *Acer caesium.*

(6) *Aphidius avenae* Holiday (XIII, Kmr.)

Host: *Chaetosiphon glaber* on *Rosa* sp., *Macrosiphum* sp. on *R. macrophylla; Myzaphis turanica* on *Rosa* sp.

(7) *Aphidius cingulatus* Rathe (LXXV, Dec.)

Host:*Pterocomma populeum* on *Populus aurea.*

(8) *Aphidius colemani* Viereck (XVI, LXXV, Gul., Srg.)

Host: *Aphis nerii* on *Calotropis; Melanaphis donacis* on *Arundo donax.*

(9) *Aphidius eglanteriae* Haliday (XIII, Kmr.)

Host: *C. tetrarhodum* on *R. macrophylla; M. rosarum* on *R. webbiana.*

(10) *Aphidius eglanteriae* Haliday (XIII, XVII, XVIII, Kmr.)

Host: *Acyrthosiphon pisum* on *Medicago* sp., *Pisum sativum* and *Trifolium* sp; *Acyrthosiphon* sp. on *Vicia sativa; M. carnosum (= evansi)* on *U. dioica.*

(11) *Aphidius gifuensis* Ashmead (XIII, XX, Kmr.)

Host: *Aphis affinis* on *M. longifolia; Aphis* sp. on *Veronica persicae and Vicia sativa.*

(12) *Aphidius hortensis* Marshall (LXXV, Kmr.)

Host: *Liosomaphis himalayensis* on *Berberis pseudoumbulata.*

(13) *Aphidius matricariae* Haliday (XIII, XXI, XX, LXXIII, LXXV, Jmu., Man., Kmr. Kup.)

Hosts: *A. craccivora* on *Solanum nigrum; A. fabae* on *Rumex nepalensis; Aphis* sp. on *Cynoglossum glochidatum; B. helichrysi* on *Ageratum conyzoides; C. elaeagini* on *carduns edelbergi; Chaitophorus kapuri* on *Populus* sp., *Phorodon cannabis (=Diphordon cannabis)* on *Cannabis saliva; Microlophium carnosum* on *U. dioica; M. persicae* on *Descuraina sophia; Lycopersicon esculentum, Phlox dummondii; Myzus umecola* on *Rubia cardifolia; P. cannabis* on *C. sativa.*

(14) *Aphidius salicis* Haliday (XIII, XVIII, Srg., Kmr.)

Host: *Aphis* sp. on *Dioscorea deltoidea, Seneio chrysanthemodes, Caveriella biswasi* on *Ferula jacschkeana; Semiaphis heraclei* on *Daucus carota.*

(15) *Aphidius salignae* Watanabe (XXI, Ban.)

Host: *Tuberolachnus salignus* on *Salix alba.*

(16) *Aphidius smithi* Subba Rao & Sharma (XIII, XVI, XVII, LXXV, Kmr.)

Host: *A. pisum* on *Lathyrus aphaca; Medicago* sp., *Ranunculus muricatus* and *Ranunculus* sp.

(17) *Aphidius transcaspious* Telenga (XXI, Poo.)

Host: *Aphis* sp. on *Prunus domestica*

(18) *Aphidius uzbekistanicus* Luzhetzki (XII, Kmr.)

Hosts: *Impatientinum impaticus dalhousiensis* on *Impatiens; Sitobium* sp. on *P.* flaccidum.

(19) *Aphidius* spp. (XII, XIII, XV, XVI, XX, LXXV, Kmr., Jmu., Gul., Zab., Srg., Band., Bee.)

Hosts: *Aphis farinosa* on *S. aegyptica, S. alba* and *Salix* sp., *Capitophorus* sp. on *Carduus* sp., *Chaetosiphon gracilicorne* on *Rosa webbiana; Chaetosiphon* sp. on *Rosa* sp., *Chaitophorus himalayensis* on *Salix flabellaris; Coloradoa rufamaculate* on *Chrysanthemum* sp., *Eriosoma* sp. on *Ulmus* sp., *Eumyzus* on *Nepeta cataria; Lachnus tropicalis* on *Pinus wallichiana; M. sanborni* on *Chrysanthemum morifollum* and *Chrysanthemum* sp., *Myzus* sp. on *Euphorbia helicoscopia* and *Sisymbrium; Rhopalosiphum nymphaeae* on *Nymphaea alba.*

(20) *Aeropraon lepelleyi* (Waterston) (VI, VIII, XV, Kmr.)

Hosts: *Eriosoma (Schizoneura) ulmi* on *Ulmus villosa (=laevigata); Schizoneurella indica* on *Ulmus wallichiana.*

(21) *Binodoxys kashmirensis* Takada & Rishi (LXXV, Har.)

Host: *Macrosiphum* sp. on *Rosa brunonii.*

(22) *Binodoxys uroleucon* Takada & Rishi (LXXV, Srg.)

Host: *Uroleucon* sp. on *C. morifolium.*

(23) *Diaeretiella rapae* (M'Intosh) (XII, XIII, XVIII, XXI, LXXV, Srg., Kmr., Raj.)
Hosts: *A. craccivora* on *Solanum tuberosum, B. brassicae* on *Brassica oleraceae* var. *acephala, D. sophia; Lipaphis erysimi* on *Brassica napus (=campestris); Uroleucon sonchi* on *Sonchus oleraceous.*

(24) *Ephedrus campestris* Stary (LXVIII, Kmr.)
Host: *M. sanborni.*

(25) *Ephedrus (E.) dioscoreae* Bhagat (XI, XX, Kmr.)
Host: *Aphis* sp., *Aulacorthum* sp. on *D. deltoidea.*

(26) *Ephedrus minor* Stelfox (LXXV, Har.)
Host: *Cavariella* sp. on *S. alba.*

(27) *Ephedrus niger* (XVIII, LXXV, Kmr.)
Hosts: *Macrosiphoniella* sp. on *Chrysanthemum* sp., *Urolcucon* sp. on *C. morifolium.*

(28) *Ephedrus persicae* Froggatt (XIII, XV, XVII, LXXIII, Kmr.)
Hosts: *A. craccivora* on *Rumex acetosella; C. populeti* on *P. caspica; H. pruni* on *Prunus domestia insititia; M. donacis* on *A. donax.*

(29) *Ephedrus plagiator* (Nees) (VIII, XIII, XV, Kmr. Duc.)
Host: *Procipuilus* sp. on *L. quinquelocularis; Tetraneura* sp. on *Ailantus altissma.*

(30) *Ephedrus salicicola* Takada (XIII, XV, Kan., Kmr.)
Host: *Cavariella aegopodi* on *S. aegyptica; Cavariella* sp. on *S. balylonica; Eumyzus* on *Nepeta catariai.*

(31) *Ephednes srinagarensis* Stary & Bhagat (VIII, XIII, LXXIII, Kmr.)
Host: *Prociphilus* sp. on *L. quinquclocalaris.*

(32) *Ephedrus (E.) urticae* Bhagat (XI, Bee.)
Host: *M. carnosum* on *U. dioica.*

(33) *Ephedrus* (*Ephedrus*) sp. (XXI, Ban., Jmu.)
Hosts: *Acyrthosiphon* sp. on *Quercus elex; Aulacorthum* sp. on *Spirea* sp.

(34) *Kashmiria aphidis* Stary & Bhagat (VIII, LXXIII), Per.-m)
Host: *Myzus* sp., *Prociphilus* sp. on *L. quinqueclocularis.*

(35) *Lipdexis gracilis* Forster (XIII, XVI, XVIII, XXI, LXXV, Kmr., Jmu.)
Hosts: *Acyrthosiphon rubi* on *Rubus fruticosus; A. affinis* on *M. aquatica* and *M. longifolia (=sylvestris); A. craccivora* on *R. pseudoacacia* and *S. nigum; A. gossypii* on *Z. elegans; A. nerii* on *V. thapsus; L. himalayensis* on *B. pseudoambulata.*

(36) *Lipolexis scutellaris* Mackauer (LXXIII, Kmr.)
Host: *A. craccivora* on *R. pseudoacacia.*

(37) *Lysaphidius staryi* Bhagat (IX, Srg.)
Host: *Macrosiphum (Sitobion) avenae* on *Pennisetum flaccidum.*

(38) *Lysiphlebia mirazai* Shuja-uddin (IV, Kmr.)
Host: *Sitobion avenae (=Macrosiphum avenae)* on *Triticum vulgare.*

(39) *Lysiphlebus ambiguous* Haliday (XVIII, Kmr.)
Host: *Sipha (Rungsia) maydis* on *S. halepense.*

(40) *Lysiphlebus confusus* Trembly & Eady (XIII, XVIII, Kmr.)
Hosts: *Aphis* nr. *intybi* on *Cichorium intybus; Chaitophorus niger* on *P. caspica; S. maydis* on *Hordeum vulgare, S. halepense* and *S. nilgare; Melanaphis sacchari* on *Sorghum* sp.

(41) *Lysiphelebus (Phelbus) fabarum* (Marshall) (XIII, XVI, Kmr.)
Hosts: *A. rubi* Narzikulov on *R. fruticosus; A. craccivora* on *R. nepalensis; Aphis* nr. *intybi* on *Cichorium intybus; Aphis* nr. *salviae* on *Salvia moorcraftiana.*

(42) *Monoctonus crepidis* Haliday (XIII, Srg.)
Hosts: *Liosomaphis atra* on *Berberis psuendoumballatus; Liosomaphis* sp. on *B. zebbiliana.*

(43) *Monoctonus (Monoctonus) fotedari* Bhagat (V, Kmr.)
Host: *M. rosarum* on *Rosa* sp.

(44) *Monoctonus* sp. (LXXV, Kmr.)
Host: *Macrosiphoniella oblonga* on *Artemisia vestita.*

(45) *Pauesia (Kashmirpausia) hazratbalensis* Bhagat (VII, Srg.)
Host: *Lachnus* sp. on *Thuja orientalis(=Biota orientalis)*

(46) *Pauesia (Paraphidius) gulmargensis* Bhagat (VII, Gul.)
Host: *Cinara lachnirostris* on *Pinus* sp.; *Cinava* sp. on *Abies pindrow*

(47) *Pauesia (Paraphidius) himalayensis* Bhagat (VII, XV, Srg. Zab.)
Host: *L. tropicalis* on *P. wallichiana.*

(48) *Pauesia (Paraphidius) pini* (Haliday) (VII, XV, Gul., Tang., Bar.)
Host: *C. lachnirostris* and *Cinara* sp. on *P. wallichiana.*

(49) *Praon abjectum* Haliday (XIII, LV, LXXIII, Srg. Kmr., Kok.)
Hosts: *A. farinosa* on *S. alba* and *S. babylonica; Aphis grossulariae* nr. *pollinosa* on *E. hirsutum; M. donacis* on *A. donax; Myzaphis rosarum* on *Rosa* sp.

(50) *Proan dorsale* Haliday (XIII, XVIII, XX, Kmr.)
Hosts: *M. sanborni* on *C. morifolium; Macrosiphoniella* sp. on *Chrysanthemum* sp.

(51) *Proan kashmirensis* Bhagat (XI, XV, Kmr.)
Hosts: *A. farinosa* on *S. alba; Chaitophorus pakistanicus* on *Salix* sp.

(52) *Praon nymphaeae* Subba Rao & Sharma (XIII, XVIII, Srg.)
Hosts: *R. nymphaeae* on *Nymphaea* spp.

(53) *Praon Stagona* Takada & Rishi (XVI, LXXIII, Zab.)
Hosts: *Amphicercidus tuberculatus* and *Stagona* sp. on *L. quinquelocularis.*

(54) *Praon volucre* Haliday (XII, XIII, Nag, Nis, Srg.)
Hosts: *Aphis* sp. on *Lespadeza* sp.; *Hyalopterus pruni (=arundis)* on *P. domestica* and *B. pseudoumiellatus; Macrosiphum* sp. on *Rosa* sp.; *M. turanica* on *Rosa* sp.; *Uroleucon* sp. on *S. asper.*

(55) *Proan* sp. (XIII, XVIII, Kmr.)

Host: *Macrosiphoniella* sp. on *Chrysanthemum* sp.

(56) *Toxares deltiger* Westwood (XIII, XV, XVII, XX, Srg., Kmr., Gul.).

Hosts: *A. craccivora* on *R. pseudoacacia; Betacallis* on *C. maculatum; Eumyzus* sp. on *N. cataria; H. pruni* on *P. domestica insititia; Macrosiphoniella yomogifoliae* on *Artemisia alsinthium; Ovatus* nr. *crataegarius* on *M. longifolia.*

(57) *Toxares macrosiphophagum* Shuja-uddin (LXVI, Kmr.)

Host: *M. sanborni* on unknown host plant.

(58) *Toxares shigai* Takada (XXVII, LXXV, Har., Srg. Kmr.)

Host: *A. farinose* on *Salix alba.*

(59) *Toxares zakai* Shuja-uddin (LXVI, Kmr.)

Host: *M. persicae* on *S. nigrum.*

(60) *Trioxys (Betuloxys) intermediatus* (LXIV, Sop.)

Host: Unknown aphid species on *Centaurea iberica.*

(61) *Trioxys (Binodoxys) acalephae* (Marshall) (XIII, XX, LXXIII, Bur., Zab., Srg. Kmr.)

Hosts: *A. affinis* on *Mentha longifolia; Macrosiphum* sp. on *Rosa* sp.; *Phorodon cannabis (=Diphorodon cannabis)* on *C. sativa.*

(62) *Trioxys (B.) basicurvus* Shuja-uddin (LXV, Sop.)

Host: *A. gossypii* on *Rubus* sp.

(63) *Trioxys (B.) brevicornis* Haliday (XIII, Srg.)

Host: *C. pakistanicus* on *S. babylonica.*

(64) *Trioxys (Binodoxys) centaureae* Haliday (XIII, XVIII, XX, LXXIII, Srg., Kmr.)

Hosts: *M. artimisae* and *M. pseudoatemisae* on *Artemisia India; M. sanborni* on *C. morifolium; Macrosiphenielia* sp. on *A. millefolium* and *Chrysanthemum* sp.

(65) *Trioxys (B.) indicus* Subba Rao & Sharma (XIII, XV, XVII, XX, XXI, LXXIII, Kmr., Khi.)

Hosts: *A. affinis* on *M. longifolia; A. craccivora* on *R. pseudoaccia; R. nepalensis; S. miniatum* and *S. nigrum; A. f. solanella* on *R. nepalensis; A. gossypii* on *Melia azedarach* and *U. dioica; A. pomi* on *M. Pumila (= M. domestica); Eriosoma* sp. on *Cotoneaster aitkinsoni; Macrospheniella pseudoartemisae* on *Artemisia absinthium; Macrosiphum* sp. on *Rosa brunonii; Mindarus japonicus* on *Abies pindrow; Phorodon (Paraphorodon)* sp. on *Punica granata.*

(66) *Trioxys (B.) jaii* Bhagat (X, XV, Srg.)

Hosts: *A. craccivora* on *R. pseudoacacia; Shivaphis celti* on *C. australis.*

(67) *Trioxys (B.) longispinus* Shuja-uddin (LXIX, Kmr.)

Host: *A. farinosa* on *S. caprea.*

(68) *Trioxys (B.) rosaecola* Bhagat (X, Kmr.)

Host: *M. rosarum* on *R. webbiana.*

(69) *Trioxys (B.) rubicola* Shuja-uddin (LXV, Sop.)
Host: *A. gossypii* on *Rubus* sp.

(70) *Trioxys (B.)* sp. (XV, XXI, Kmr., Ban.)
Hosts: *A. craccivora* on *Robinia* sp.; *M. japonicus* on *A. pindrow.*

(71) *Trioxys (Trioxys) complanatus* Quilis (XIII, XVIII, Srg., Kmr.)
Hosts: *A. pisum* on *L. aphaca; Therioaphis* sp. on *M. polymorpha; M. sativa* and *V. sativa.*

(72) *Trioxys (Trioxys) pannonicus* Stary (XIII, XX, Kmr.)
Hosts: *M. pseudoartemisae* on *Artemisia* spp., *A. absinthium* and *A. indica.*

(73) *Trioxys (T.) rishii* Stary & Bhagat (XIII, XV, XVII, LXXIII, Srg.Kmr.)
Hosts: *A. craccivora* on *R. pseudoacacia; Brachycaudus* sp. on *C. edelbergi; Capitophorus* sp. on *Carduus* sp., *H. pruni* on *P. d. insititia.*

(74) *Trioxys (T). shivaphis* Takada (X, XV, Srg.)
Host: *S. celti* on *C. australis.*

(b) Family: Braconidae

(1) *Apanteles conspersae* (II, III, Kmr.)
Host: *L. obfuscata* on *populus* spp. and *salix* spp.

(2) *Apanteles flavicoxis* Marsh (LXII, Kmr.)
Host: *L. obfuscata* on *salix* spp.

(3) *Apanteles indiansis* Marsh (LXII, Kmr.)
Host: *L. obfuscata* on *salix* sp.

(4) *Apanteles liparidis* Bouche (II, III, LXII, Kmr.)
Host: *L. obfuscata* on *Salix* spp. and *Populus* spp.

(5) *Apanteles melanoscelus* (Ratzburg) (LXII, Kmr.)
Host: *L. obfuscata* on *Salix* spp. and *Populus* spp.

(6) *Apanteles porthetriae* Muesebeck (LXII, LXXVIII, Kmr.)
Host: *L. obfuscata* on *Salix* spp. and *Populus* spp.

(7) *Apanteles nr. solitarius* (LXXVIII, Kmr.)
Host: *L. obfuscata* on unidentified plant

(8) *Apanteles* spp. (III, LVII, Sgr. Ldk.)
Host: *Archips subsidiaria* on *M. pumila, Ostrinia kashmirica (*Corn borer*)* on *Cirsium (= Cnicus* sp.), *L. obfuscata*on *Salix* spp. *Populus* spp.

(9) *Blacus (Blacus) imitator* Papp. (LVI, Sgr.)
Host: unknown insect and plant species

(10) *Bracon brevicornis* (III, Ach. Lark.)
Host: *O. kashmirica*

(11) *Phaenocarpa kashmirensis* Bhat (XXIII, Khi, Pah. Son.)
Host: Unknown host insect and plant species.

(12) *Phaeocarpa tridenta* Bhat (XXIII, Gul, Son, Pah)
Host: unknown insect and plant species.

(13) *Protatus* n. gen. (LXII, Kmu.)
Host: *Sinnoxyla* sp. On *Dalbergia* sissou

(14) *Spathius* sp. (XXV, LV, Kmr.)
Host: *Scotylus* sp. on *M. pumila.*

(c) Family: Ichneumonidae

(1) *Anilastus* sp. (III, Kmr.)
Host: *L. obfuscata* on *Salix* spp., *Populus* spp.

(2) *Bassus lactatorius* (Moley) (I, Kmr.) (=*Ichneumon lactatorius)*
Host: Unknown insect and host plant species.

(3) *Campoplex longipes* Moley. (I, Kmr.)
Host: Unknown insect and host plant species.

(4) *Cephalobolus parvipes* Moley (I, Kmr.)
Host: Unknown insect and plant species.

(5) *Coccygomimus* sp. (III, Sgr.)
Host: *L. obfuscata* on unknown host plant species.

(6) *Cryptus nursei* Cameron (XXVI, Kmr.)
Host: Unknown insect species.

(7) *Deuteroxorides indicus* Gupta (XXXI, Pah.
Host: Coleopteran larvae on hardwood coniferous tree)

(8) *Ecthrus coracinus* Gupta (XXIX, Gul.)
Host: Coleopteran larvae (wood-boring)

(9) *Ephonites ruficornis* Cameron (XXVI, Kmr.)
Host: Unknown host insect and plant species.

(10) *Eridolius aithogaster* Gupta (XXXV, LXXIV, Pah.)
Host: Larvae of sawflies (Tenthredinidae)

(11) *Exeristes roborator* (III, Lar., Ach.)
Host: *O. kashmirica (Pyralidae)* on *Cirsium* sp. (= *cricus* sp.)

(12) *Exetastes fornicator* (Moley) (I, Kmr.)
(= *Ichneumon fornicator*
Host: Unknown insect and plant species

(13) *Gasteruption kashmirensis* Cameron (XXVI, Kmr.)
Host: Unknown insect and plant species

(14) *Horogenes* sp. (LXX, Sgr., Kmr.)
Host: *Plutella xylostella* (= *maculipennis*) on unknown crop plant.

(15) *Inocellia fulvostigmata* Horstmann (XXXXI, Kis.)
Host: Unknown insect and plant species.

(16) *Itoplectis himalayansis* Gupta (XXX, LXII, Pah. Bat. Srg.)
Host: *L. obfuscate* on *Salix* spp. and *Populus* spp.

(17) *Itoplectis tibetensis* Perking (XXX, Gul.)
Host: Lepidopterous pupal on herbaccous bushes along forest-edge.

(18) *Itoplectis* sp. (LXX Sgr.)
Host: *P. xylostella* on unknown crop

(19) *Paniscus montanes* Cameron (XXVI, Kmr.)
Host: Unknown insect and plant species.

(20) *Paniscus ocellaris* Moley (I, Sgr.)
Host: Unknown insect and plant species.

(21) *Paniscus kashmirensis* Cameron (XXVI, Kmr.)
Host: Unknown (Insect and plant).

(22) *Pimpla archica* Zeltenh. (I, Kmr.)
Host: Unknown insect and plant species.

(23) *Pimpla alternans* (III, Kmr.) (= *Itoplectis alternans*)
Host: Unknown plant species.

(24) *Pimpla* sp. (II, XXXXVI, LXXVIII, Kmr.)
Host: Pupae of *L. obfuscata* on *Salix*. spp. and *Populus* spp.

(25) *Podoschisthus scutellaris* Designes (XXXI, Kmr.)
Host: Larvae of woodboring (hardwood) coleopterans.

(26) *Poemenia kashmirica* Gupta (XXXI, XXXIII, Pah. Kmr.)
Host: Larvae of coleopterans, wood boring (coniferous)

(27) *Polyblastus kashmirensis* Gupta (XXXIV, Pah.)
Host: Larvae of sawfly on unknown plant species.

(28) *Pristomerus* sp. (III, Lar., Ach.)
Host: *Ostrinia kashmirica* on *Crisium* sp.)

(29) *Promethes albiventralis* Diller (XXVIII, Ldk.)
Host: Unknown insect and plant species.

(30) *Scambus* (*Ateleophadnus*) *aithomelus* Gupta & Tikar (XXXVII, Sgr.)
Host: *O. kashmirica on Crisium wallichi)*

(31) *Scambus flavicrus* (III, Lar. Ach.)
Host: *O. kashmirica* on *Crisium* sp. (= *Crinus* sp.)

(32) *Scambus (S.) indicus* Gupta & Tikar (XXXVII, Gul.)
Host: Larvae of Lepidopterans on unknown host plant.

(33) *Scambus (S.) kashmiricus* Gupta & Tikar (XXXVII, Sho.)
Host: Lep. Larvae and spider eggs on unknown host plant species.

(34) *Scambus nursei* (Cameron) (LIV, Kmr.)
Host: Lep., Col. and Larvae on unknown host plant species.

(35) *Theronia atalantae atalantae* (III, XXXXVI, LXII, Kmr.)
Host: *L. obfuscata* on *Salix* spp. and *Populus* spp.

(36) *Theronia* sp. (II, XXXXV, LXXVIII, Kmr.)
Host: *L. obfuscata* (pupal parasite) on *Salix* spp. and *Populus* spp.

(37) *Xorides* (*Xorides*) *anthracinus* Gupta & Chandra (XXXVI, Pah.)
Host: Unknown insect and plant species.

(38) *Xorides* (*X.*) *minimus* Gupta & Chandra (XXXVI, Gul.)
Host: Unknown insect and plant species.

Key to the Abbreviations and Roman numbers (I-LXXVIII)

Ach.- Achabal, Ana.- Anantnag, Ban.- Banihal, Band.- Bandipore, Bar.- Baramulla, Bat.-Batote, Bee.- Beerwah, Bur.- Burzhama, Che.- Chesmashi, Dac.- Dachigam, Duc.- Ducksum, Gan.- Ganderbal,Gul.- Gulmarg, Han.- Handwara, Har.- Harvan, Hari.- Hariparvat, Hee.- Heewan, J&K.- Jammu & Kashmir, Jmu.- Jammu, Kan.- Kangan, Khi.- Khilanmarg, Kis.- Kishtwar, Kmr.- Kashmir, Kok.- Kokernag, Kup.- Kupwara, Lar.- Larkipora, Ldk.- Ladakh, Man.- Manasbal, Nag.- Nagin, Nis.- Nishat, Pah.- Pahalgam, Pat.- Pattan, Peri.m-. Peri Mahal, Poo.- Poonch, Raj.- Rajouri, Sho.- Shopian, Son.- Sonmarg, Sop.- Sopore, Srg.- Srinagar, Zab.- Zabervan.

I= Ahmad & Dar(1978), II = Anonymous (1969), III = Anonymous (1970), IV = Bhagat (2002), V = Bhagat (1981a), VI = Bhagat (1981b) VII = Bhagat (1981c), VIII = Bhagat (1982), IX = Bhagat (1982*a*), X = Bhagat (1982*b*), XI = Bhagat (1982*c*), XII = Bhagat (1983), XIII = Bhagat (1984), XIV = Bhagat (1985), XV = Bhagat (1985*a*), XVI = Bhagat (1986), XVII = Bhagat (1986*a*), XVIII = Bhagat (1986*b*), XIX = Bhagat (1987), XX = Bhagat (1987*a*), XXI = Bhagat & Ahmad (1995) XXII = Bhagat *et al.* (1992), XXIII = Bhat (1979), XXIV = Boucek *et al.* (1978), XXIVA = Brunetti(1920), XXV = Buhroo *et al.* (2001), XXVI = Cameron (1906), XXVII = Dharmadhikari & Ramaseshia (1970), XXVIII = Diller (1984), XXIX = Gupta (1978), XXX = Gupta (1967), XXXI = Gupta (1980), XXXII = Gupta (1984), XXXIII = Gupta (1985), XXXIV = Gupta (1985*a*), XXXV = Gupta (1991), XXXVI = Gupta & Chandra (1974), XXXVII = Gupta & Tikar (1967), XXXVIII = Hansson (1985), XXXIX = Hayat (1972), XXXX = Hayat (1975), XXXXI = Horstmann (1992), XXXXII = Husain & Agarwal (1981), XXXXIII = Mani & Kaul (1973), XXXXIV = Masoodi & Trali (1987), XXXXV = Masoodi *et al.* (1986), XXXXVI = Masoodi *et al.* (1986*a*), XXXXVII = Masoodi *et al.* (1989), XXXXVIII = Masoodi *et al.* (1989a), IL = Mathur (1956),ILA = Mathur & Sharma (1962), L = Moley (1913), LI = Narayanan (1957), LII = Narayanan (1961), LIII = Narayanan & Lal (1959), LIV = Narayanan & Lal (1955), LV = Narendran *et al.* (2001), LVI = Papp (1985), LVII = Pawar & Parry (1989), LVIII = Pawar *et al.* (1981), LIX = Ramaseshia & Dharmadhikari (1969), LX = Rao (1966), LXI = Rishi & Lone (1984), LXII = Rishi & Shah (1985), LXIII = Sharma & Gupta (1985), LXIIIA =Sewell & Roonwal(1965), LXIV = Shuja-uddin (1970), LXV = Shuja-uddin (1974), LXVI = Shuja-uddin (1976), LXVII = Shuja-uddin (1978), LXVIII = Shuja-uddin (1978*a*), LXIX = Shuja-uddin (1983), LXX = Simmonds & Rao (1960), LXXI – Singh & Udaya (1987), LXXII = Singh *et al.* (1984), LXXIII = Stary & Bhagat (1978), LXXIV = Rao & Hayat (1986), LXXV = Takada & Rishi (1980), LXXVI = Zaka-ur-Rab (1974), LXXVII = Zeya & Hayat (1995) and LXXVIII = Zutshi (1967).

8.3. CHECKLIST OF HOST-PLANT, INSECT AND ENTOMOPHAGOUS PARASITIC INSECT SPECIES

(Keys to the abbreviations given at the end of the Checklist)

(1) *Abies pindrow* (The Himalayan Silver Fir) (Pinaceae)

Aph. - *Cinara* sp.

Apd. – *P. (P.) gulmargensis*

Aph. – *M. japonicus* (Balsam aphid)

Apd. – *T. (B.) indicus, Trioxys* (*B.*) sp.

(2) *Acer caesium* (Mapple) (Aceraceae)

Aph. – *D. plantanoids* (Sycamose aphid)

Apl. – *Aphelinus* nr. *fulvus*

Aph. – *P. vandenboschii*

Apd. – *A. areolatus*

(3) *Achillea millefolium* (Milfoil, Yarrow, Sanguinary) (Asteraceae)

Aph. – *Macrosiphoniella* sp.

Apd. – *T. (B). centaureae*

(4) *Aesculus indicus* (Indian Horse-Cheshnut (Hippocastanaceae)

Aph. – *P. aesculi*

Apd. – *A. areolatus*, Pte. – *Asaphes suspensus* (hyperparastic)

(5) *Ageratum conyzoide* (Goat weed) (Asteraceae)

Aph. – *B. helichrysi*

Apd. – *A. matricariae*

(6) *Ailantus altissma* (Ailanto, Tree of Heaven) (Simaroubaceae)

Aph. – *Tetraneura* sp.

Apd. – *E. plagiator*

***Artemisia* spp. (Asteraceae)**

(7) *Artemisia absinthium* (Absinthe, Wormwood, Madderwood)

Aph. – *M. nr. artemisae*

Apd. – *T. (T.) pannonicus*

Aph. – *M. pseudoartemisae*

Apd. – *T. (B.) indicus*

Aph. – *M. nr. yomogifoliae*

Apd. – *T. deltiger*

(8) *Artemisia indica*

Aph. – *M. artemisae*

Apd. – *T. (B.) centaureae*

Aph. – *M. pseudoartemisae*

Apd. – *T. (B.) centaureae, T. (B.) indicus, T. (T.) pannonicus*

(9) *Artemisia vestita*

Aph. – *M. oblonga*

Apd. – *Monoctonus* sp.

(10) *Arundo donax* (Great Reed, Spanish Cane) (Poaceae)

Aph. – *M. donacis*

Apd. – *A. colemani, E. persicae, P. objectum*

***Berberis* spp. (Berberidaceae)**

(11) *Berberis pseudoumbellatus* (Barberry)

Aph. – *Liosomaphis atra*

Apd. – *M. crepidis, P. volucre*

Aph. – *L. himalayensis*

Apd. *A. hortensis, L. gracilis*

(12) *Berberis zebbiliana* (Barberry)

Aph. – *Liosomaphis* sp.

Apd.- *M. crepidis*

***Brassica* spp. (Brassicaceae)**

(13) *Brassica napus* (Rape) *B.* n. var. *glavica* (Indian Colza) (=*Brassica campestris*)

Aph. – *L erysimi* (Turnip/Mustard aphid)

Apd. – *D. rapae*

(14) *Brassica olercea var. acephala* (Knol- Khol, Turnip Kale)

Aph. – *B. brassicae*

Apd. – *D. rapae,* Pte. *A. suspensus* (hyperparasite)

(15) *Calotrophis* sp. (Asclepiadaceae)

Aph. – *A. nerii* (Oleander aphid)

Apd. – *A colemani*

(16) *Cannabis sativa* (Hemp.) (Marijuana) (Cannabaceae)

Aph. – *Phorodon cannabis*

Apd. – *A. matricariae, T. acalephae*

(17) *Capsella bursa pastoris* (Shepherd's purse) (Brassicaceae)

Aph. – *Aphis fabae* (Bean aphid)

Apd. – *A. gifuensis,* Pte.- *Pachyneuron aphidis* (hyperparasite)

(18) *Carduus edelbergii* (Thistle) (Asteraceae)

Aph. – *C. elaeagni*

Apd. – *A. matricariae, Pte. – A. suspensus,P. aphidis (hyperparasite);*

Apd. – *T. rishii, Pte.— P. groenlandicum,*

Aph. – *Capitophorous* sp.

Apd. – *Aphidius* sp., *T.(T.) rishii*

Aph. – *Brachycaudus* sp. (Thistle aphid)

Apd. – *T. (T.) rishii*

(19) *Celtis australis* (Ulmaceae)

Sc-i. – *Pulvinaria inconspiqua* (Woolly/Cottony Scale insect)

Apl. – *A. mali*

Aph. – *Shivaphis celti*

Apd. – *T. (B.) jaii, T. (T.) shivaphis*

(20) *Centaurea iberica* (Asteraceae)

Aph. – Undetermined aphid species

Apd. – *T. (B.) intermediatus*

***Chrysanthemum* spp. (Asteraceae)**

(21) *Chrysanthemum morifolium*

Aph. – *M. sanborni*

Apd. – *Aphidius* sp., *P. dorsale, T. centaureae*

Aph. – *Uroleucon* sp.

Apd. – *B. uroleucon, E. niger*

(22) *Chrysanthemum leucanthemum* (Ox-eye-Daisy)

Aph. – *U. tanaceti*

Apd. – *T. shigai*

(23) *Chrysanthemum* sp.

Aph. – *Coloradoa rufamaculata*

Apd. – *Aphidius* sp.

Aph. – *M. sanborni*

Apd. – *Aphidius* sp.

Aph. – *Macrosiphoniella* sp.

Apd. – *A. absinthi, E. niger, P. dorsale. Proan sp., T. centureae*

(24) *Cichorium intybus* (Chicory/Wild Endive) (Asteraceae)

Aph. – *Aphis* nr. *intybi*

Apd. – *L. fabarum, L. (P.) confuscum*

(25) *Cirsium* ? *wallichi* (Thistle Plant) (=*Cnicus wallichi*)

Pyr. –*Ostrinia Kashmirica* (Corn Borer)

Ich. – *Scambus (A.) aithomelus)*

(26) *Cirsium sp* (=*Cnicus* sp.)

Pyr. – *O. kashmirica*

Bra. – *Apanteles* sp., *Bracon brericornis*; Ich. – *Pristomerus* sp., *Scambus flavicrus.*

(27) *Conium maculatum* (Poison Hemlock) (Apiaceae)

Aph. – *Betacallis* sp.

Apd. – *T. deltiger*, Cyn.– *Alloxysta* sp. (hyperparasite)

(28) *Cotoneaster aitkinsonii* (Rosaceae)

Aph. – *Eriosoma* sp.

Apd. – *T. (B.) indicus*

(29) *Cynoglossum wallichi* (Boraginaceae) (=*Cynoglossum glochidatum*)

Aph. – *Aphis* sp.

Apd. *A. matricariae*

(30) *Dalbergia sissoo* (Sissoo/Sheesham tree) (Fabaceae) *? Sinoxyle* sp.

Bra. – *Protatus*

(31) *Daucus carota* (Wild carrot) (Apiaceae)

Aph. – *Semiaphis heraclei*

Apd. – *A. salicis*

(32) *Daphne oleoides* (Spurge) (Thymelaceaceae)

Aph. – *A. kurusawai*

Apd. – A *mali*

(33) *Descuraina sophia* (Flax/Flixweed)

Aph. – *B. brassicae*

Apd. – *D. rapae*

Aph. – *M. persicae* (Green Peach-potato aphid)

Apd. – *A. matriceriae*

(34) *Dioscorea deltoidea* (Dioscoreaceae)

Aph.– *Aphis* sp.

Apd. – *A. salicis, E. (E.) dioscoreae*

Aph. – *Aulacorthum* sp.

Apd. – *E. (E.) dioscoreae*

(35) *Epilobium hirsutum* (Hainy willow weeds) (Onagraceae)

Aph. – *A. affinis*

Apl. – *A. mali*

Aph. – *A.grossulariae* nr. *pollinosa*

Apd. – *P. objectum*, Meg. – *Dendroceros carpentri* (hyperparasite)

Aph. – *R. nymphaeae*

Apd. – *P. abjectum.*

(36) *Euphorbia helicoscopia* (Sun Spurge) (Euphorbiaceae)

Aph. – *Myzus* sp.

Apd. – *Aphidius* sp.

(37) *Ferula jaesckeana* (Apiaceae)

Aph. – *C. biswasi*

Apd. – *A. salicis*

(38) *Hordeum vulgare* (Barley) (Poaceae)

Aph. – *Sipha maydis*

Apd. – *Lysephlelus confuscus*

(39) *Impatiens* sp. (Balsum plant) (Balsaminaceae)

Aph. – *Impatientinum impatiens dalhousiensis*

Apd. – *A. uzbekistanicus*, Cyn. – *Alloxysta* sp. (hyperparasite)

(40) *Lathyrus aphaca* (Yellow Yetchling) (Fabaceae)

Aph. – *A.* pisum

Apd. – *T. (T.) complanatus, A. smithi*

(41) *Lespedeza* sp. (Lespedeza) (Fabaceae)

Aph - *Aphis* sp.

Apd. – *P. volucre*

(42) *Lonicera quinquelocularis* (Honey-suckle) (Caprifoliaceae)

Aph. – *Amphicercidus tuberculatus*

Apd. – *P. dorsale, Meg. – D. carpentri* (hyperparasite*);*

Apd. – *P. stagona*

Aph. – *Prociphilus* sp.

Apd. – *E. plagiator, E. srinagarensis*, Pte. – *Coruna* sp.- (hyperparasite);

Apd. – *K. aphidis*

Aph. – *Stagona* sp.

Apd. – *P. stagona.*

(43) *Lycopersicon esculentum* (Tomato) (Solanaceae)

Aph. – *M. persicae*

Apd. *A. matricariae*

(44) *Malus* spp. (*M.pumila, M. sylvestris, M. domestica)* (Apple) (Rosaceae)

Aph. – *Aphis pomi* (green apple aphid)

Apd. — *T. (B.) indicus and T. (T.) rishii*

Aph. – *E. lanigerum* (Woolly-apple aphid)

Apl. – *A. mali;* Pte. – *Asaphes fltcheri*

Sc-i. – *Quadraspidiotus perniciosus* (San Jose Scale)

Apl. – *Aphytis paramaculicornis, Aphytis* sp.(*proclia* group),

A.– lingnanensis, Encarsia (*Prospaltella) perniciosi*

Sco. – *Scotylus* spp. (Shot hole borer)

Bra. – *Spathius* sp.;

Eup. – *Eupelmus kashmiricus, E. valsus, E. vindex*;

Eur. – *Eurytoma morio* ;

Pte. – *Cheiropachus quadrum,Heydenia indica, Macromeseus harithus, Rhaphitelus maculatus*

Tor. – *Archips subsidaria* (Fruit-tree tortricid)

Bra. – *Apanteles* sp.

Tor. – *Cydia pomonella* (Codling Moth)

Tri. – *Trichogramma embryophagum* and *T. cacoeciae pallidum*

***Medicago* spp. (Fabaceae)**

(45) *Medicago polymorpha*

Aph. – *Therioaphis* sp. (Clover aphid)

Apd. *T. complanatus*

(46) *Medicago sativa* (Lucerne/ Alfalfa)

Aph. –*Therioaphis* sp.

Apd. — *T. complanatus*

Aph. – *Acyrthosiphon* sp.

Apd. – *A. smithi*

(47) *Medicago* sp.

Aph.– *A. pisum*

Apd. – *A. ervi, A. smithi*

(48) *Melia azedarach* (Persian Liliac)/Bead tree (Meliaceae)

Aph.- *A. gossypii* (Melon/cotton aphid)

Apd. –*Trioxys (B.) indicus*

***Mentha species* (Mint) (Lamiaceae)**

(49) *Mentha aquatica* (Marsh/water Mint)

Aph. – *Aphis affinis*

Apd. – *L. gracilis*

(50) *Mentha longifolia* (Horse-mint) (= *Mentha sylvestris)*

Aph. – *A. affinis*

Apd.-— *A. gifuensis, L. gracilis, Trioxys acalephae*; *Trioxys* (*B.*) *indicus*, Pte. – *A. suspensus* (hyperparasite)

Aph. – *Ovatus* nr. *crataegarius*

Apd. – *Aphidius* sp.,Enc. – *Aphidencrytus aphidivorus* (hyperparasite),*Toxares deltiger*

***Nepeta species* (Lamiaceae)**

(51) *Nepeta cataria* (Cat Mint)

Aph. – *Eumyzus* sp.

Apd. – *Aphidius* sp., *E. salicicola* and *T. deltiger*

(52) *Nepeta salvifolia*

Aph. – *Aphis* sp.

Apd. – *A. gifuensis,* Cyn. – *Phaenoglyphis* sp. (hyperparasite)

***Nymphaea* spp.** (Nymphaeaceae)

(53) *Nymphaea alba* (White water-lily)

Aph . – *Rhopalosiphum nymphaeae* (Water lily aphid)

Apd. – *Aphidius* sp.

(54) *Nymphaea* sp.

Aph. – *R. nymphaeae*

Apd. – *P. nymphaeae*

(55) *Pennisetum flaccidum* (Poaceae)

Aph. – *Sitobiom avenae* (=*Macrosiphum avenae*) (Grain aphid)

Apl. – *A. mali;*

Apd. – *A. uzbekistanicus, Cyn.* – *Alloxysta*— (hyperparasite);

Apd. —*Lysaphidius staryi*

Aph. – *Sitobion* sp.

Apd. – *A. uzbekistanicus.*

(56) *Phlox dummondi* (Polemoniacae)

Aph. – *M. persicae*

Apd. – *A. matricariae*

(57) *Phragmites communis* (Common Reed) (Poaceae)

Aph. – H. *pruni*

Apd. – *P. volucre*

Pinus spp. (Pinaceae)

(58) *Pinus roxburghii* (Chir Pine/The Himalayan long-leaved Pine)

Aph. – *C. lachnirostris* (Black Pine Aphid)

Apd. – *P. (P.) pini*

(59) *Pinus wallichina* (= *excelsa*) (Blue Pine/Bhutan Pine)

Aph. – *C. lachnirostris*

Apd. – *P. (P.)* pini

Aph. – *Cinara* sp.

Apd. *P.* —(*P.*) pini

Aph. – *L. tropicalis*

Apd. – *Aphidius* sp., *P. (P.) himalayensis*

(60) *Pinus* sp.

Aph. – *C. lachinirostris*

Apd. – *P. (P.) gulmargensis, P. (P.) pini*

Aph. – *Cinara* sp.

Apd. – *P. (P.) pini*

(61) *Pisum sativum* (Pea) (Fabaceae)

Aph. – *Acyrthosiphon pisum* (Pea aphid)

Apd. – *A. ervi* and *A. smithi*

(62) *Poa angustifolia* (Grass) (Poaceae)

Aph. – *S. maydis*

Apd. – *Lysiphleibus ambiguus*, Cyn. – *Alloxysta* sp. (hyperparasite)

***Populus species* (Salicaceae)**

(63) *Populus aurea*

Aph. – *Pterocomma populeum*

Apd. – *A cingulatus*

(64) *Populus caspica* (white Poplar)

Aph. – *Chaitophorus niger*

Apd. – *L. (P.) confusus*

Aph. – *C. populeti*

Apd. – *E. persicae*

Aph. – *Pemphigus mordvilkoi*

Apl. – *Protaphelinus nickolskajae*

(65) *Populus ciliata* (The Himalayan Poplar)

Aph. – *Epipemphigus imaicus*

Apl. – *P. nickolskajae*

Aph. – *P. aedificator*

Apl. – *P. nickolskajae*

Aph. – *P. immunis*

Apl. – *P. nickolskajae, Ene A. aphidivorus* (hyperparasite)

(66) *Populus nigra* var. *italica* (Lombardy Poplar)

Aph. – *P. bursarius*

Apl. – *P. dargahae*

Aph. – *P. immunis*

Apl. – *P. nickolskajae*

(67) *Populus* Species

Aph. – *Chaitophorus kapuri*

Apd. – *Adialytus salicaphis*; *A. matricariae*

Aph. – *P. napeus*

Apl. – *P. nickolskajae*

Lym. – *L. obfuscata* (Gypsymoth)

Bra. – *Apanteles conspersae, A. liparidis, A. melanoscelus, A. porthetriae, A. solitaria, Apanteles* sp; Cha. – *Brachymeria intermedia, A.lasus.*; Eup. – *Anastatus kashmirensis* and *Anastatus* sp., Ich. – *Auilastus* sp., *Coccygomimus* sp., *Itoplectis himalayensis, Pimpla* sp., *Theronia atlantae atlantae;*

Mus. – *Muslina stabularis*;

Sar. – *Sarcophaga* sp.,

Tac. – *Corcelia* sp., *Drino discrete, Exorista rossica*;

Torm.– *Monodontomerus acreus*

***Prunus species* (Rosaceae)**

(68) *Prunus armeniaca* (Common Apricot)

Sc-i. – *Eulecanium coryli*

Apl.- *Aphelinus* sp.

(69) *Prunus domestica* (Common plum)

Aph. – *Aphis* sp.

Apd. – *Aphidius transcaspious*

Aph. – *Hyalopterus pruni (=arundinis)*

Apd. – *P. volucre*

(70) *Prunus domestica* var. *insititia* (Cultivated variety of plum)

Aph. – *H. pruni*

Apd. – *E. persicae, P. volucre, Pte. – Coruna sp. and Pachyneuron aphidis* (hyperparasites); *T. deltiger* and *T. rishii*

(71) *Punica granatum* (Pomgernate) (Punicaceae)

Aph. – *Phorodon (=Paraphorodon)* sp.

Apd. – *T. (B.) indicus*

(72) *Quereus ilex* (Holi or Holm-Oak) (Fagaceae)

Aph. – *Acyrthosiphon* sp.

Apd. – *Ephedrus (E.)* sp.

***Ranunculus species.* (Ranunculaceae)**

(73) *Ranunculus muricatus* (Edible Buttercup)

Aph. – *Acyrthosiphon* sp.

Apd. – *A. smithi*

(74) *Ranunculus* sp.

Aph. – *Acyrthospihon* sp.

Apd. – *A. smithi*

***Rosa species.* (Rosaceae)**

(75) *Rosa brunonii* (Himalayan Musk Rose)

Aph. – *Macrosiphum* sp.

Apd. – *Aphidius* sp; Cyn. – *Alloxysta* sp. (hyperparasite);
Apd.– *T. (B.) indicus, Binodoxys kashmirensis.*

(76) *Rosa canina*

Aph. – *Chaetosiphon tetrarhodus*

Apd. – *Aphidius* sp., Euc. – *Prionomitus* sp. (hyperparasite)

(77) *Rosa chinensis* (China/ Bengal or Monthly Rose) (=*Rosa indica*)

Aph. – *Macrosiphum rosae* (Rose aphid)

Apd. – *A. avenae, Pte. – A. suspensus, Coruna* sp. (hyperparasites)

(78) *Rosa macrophylla* (Wild Rose)

Aph. – *C. chaetosiphon*

Apd. – *A. rosae, Pte. – A. suspensus; A. avenae A. suspensus* (hyperparasite)

Aph. – *C. tetrarhodum*

Apd. – *A. eglanteriae*

Aph. – *Chaetosiphon* sp.

Apd. – *Proan* sp.

Aph. – *Macrosiphun* sp.

Apd. – *Aphidius* sp.

(79) *Rosa webbiana* (Wild Rose)

Aph. – *C. gracilicorne*

Apd. – *Aphidius* sp.

Aph. – *Chaetosiphon* sp.

Apd. – *Proan* sp., Enc. – *A. aphidivorus* (hyperparasite)

Aph. – *M. rosae*

Apd. – *A. avenae* Meg. – *D. carpenteri* (hyperparasite);*T. acalephae*,
Pte. – *P. aphidis* (hyperparasite)

Aph. – *Myzaphis rosarum*

Apd. – *A. eglanteriae, T. (B.) rosaecola*

(80) *Rosa* Species

Aph. – *Chaetosiphon glaber*

Apd. – *A. avenae*

Aph. – *C. tetrarhodus*

Apd. – *Aphidius* sp., Euc. – *Prionomitus* sp. (hyperparasite)

Aph. – *Chaetosiphon* sp.

Apd. – *Aphidius* sp., Cyn. – *Alloxysta* sp. (hyperparasite), Pte.- *Coruna* sp. (hyperparasite)

Aph. – *M. rosae*

Apd. – *Aphidius* sp; Pte. – *Coruna* sp.

Aph. – *Macrosiphum* sp.

Apd. – *P. volucre*, Pte.- *A. suspensus* (hyperparasite); *T. (B.) acalephae*

Aph. – *Myzaphis rosarum*

Apd. – *M. (M.) fotedari, P. abjectum, P. volucre, Cyn.-Alloxysta* sp.(hyperparasite)

Aph. – *M. turanica*

Apd. – *Avenae* and *P. volucre*

(81) *Robinia pseudoacacia* (False Acacia/Black locrst) (Fabaceae)

Aph. – *Aphis craccivora* (Groundnut/Cowpea aphid)

Apd. – *L. gracilis, L. scutellaris, L.* (P.) *fabarum T. (B.) indicus, T. (B.) jaii, T. (Binodoxys)* sp., *T. deltiger, T. (T.) rishii;* Apl. – *A. mali*

(82) *Rubia cardifolia* (Indian Madder) (Rubiacea)

Aph. – *Myzus umecola*

Apd. – *A. matricariae*

***Rubus* spp. (Rosaceae)**

(83) *Rubus fruticosus* (Alish, Akhi, Kanachi)

Aph. – *Acyrthosiphon rubi*

Apd. -- *L. gracilis, L.* (P.) *fabarum*

(84) *Rubus* sp.

Aph. – *A. gossypii*

Apd. – *T. (B.) basicurvus, T. (B.) rubicola*

***Rumex* spp. (Ploygonaceae)**

(85) *Rumex acetosella* (Sheep Sorrel)

Aph. – *A. craccivora*

Apd. – *Ephedrus persicae*

(86) *Rumex nepalensis* (Sorrel/Broad-leaved Dock herb)

Aph. – *A. craccivora*

Apd. – *T. indicus*

Aph. – *A. fabae*

Apd. – *A. marticariae*

Aph.– *A. fabae solanella*

Apd. – *L. (P.) fabarum, T. (B.) indicus*

***Salix* spp. (Willows) (Salicaceae)**

(87) *Salix alba* (European-willow/ White-willow)

Aph. – *A. farinosa*

Apd. – *Aphidius* sp., Enc- *Prionomitus* (hyperparasite); *P. objectum, P. kashmirensis, T. shigai*

Aph. – *Chaetophorus leucomelas*

Apd. – *T. shigai*

Aph. – *Cavariella* sp.

Apd. – *E. minor*

Aph. – *Tuberolachnus salignus*

(88) *Salix aegyptica*

Aph. – *A. farinose*

Apd. – *Aphidius* sp., Pte. – *P. aphidius* (hyperparasite)

Aph. – *Cavariella aegopodi* (Willow-Carrot aphid)

Apd. – *E. salicicola*

(89) *Salix babylonica* (Weeping-Willow)

Aph. – *A. farinosa*

Apd. – *P. abjectum*

Aph. – *C. pakistanicus*

Apd. – *T. (B.) brevicornis*

Aph. – *Cavariella* sp.

Apd. – *E. salicicola*

(90) *Salix caprea* (Sallow Goat willow)

Aph. – *A. farinosa*

Apd. – *T. (B.) longispinus*

(91) *Salix flabellaris*

Aph. – *C. himalayensis*

Apd. –*Aphidius* sp.

(92) *Salix fragilis* (Crack-Willow/Kashmir Willow/ Red-wood willow)

Sc-i. – *Lepidosaphes kirgisica*

Apl. – *A. paramaculicornis*

(93) *Salix* Species

Aph. – *A. farinose*

Apd. – *Aphidius* sp.

Aph. – *C. pakistanicus*

Apd. – *P. kashmirensis*

Lip. – *Porthetria dispar* (Gypsy/Tussck Moth)

Eup. – *Anastatus kashmirensis*

Lym. – *Lymantria obfuscata* (Gypsy Moth)

Bra. – *Apanteles conspersae, A. flavicozis, A. Indiansis, A. liparidis, A. porthetriae, A. melanoscelus, A.* nr. *solitarius, Apanteles* sp.

Cha. – *Brachymeria euloeae, B. intermedia, B. lasus*;

Eup. – *Anastatus bifasciatus, A. kashmirensis, Anastatus* sp.

Eur. – *Eurytoma* sp.;

Ich. – *Anilastus* sp. ; *Coccygomimus* sp.; *Itoplectis himalayanum, Pimpla* sp.; *Theronia atalantae atlantae, Theronia* sp.;

Mus. – *Muslina stabulans*;

Sar. – *Sarcophaga* sp.;

Tac. – *Compsilura* sp.; *Corcelia* sp.; *Drino discreta D. inconpicuoides, Exorista*;

Torm. – *Monodontomerus acreus.*

(94) *Salvia moorcroftiana* (Moorcroft sage) (Lamiaceae)

Aph. – *Aphis* nr. *salviae*

Apd. – *L. (P.) falarum*

(95) *Senecio chrysanthemoides* (Asteraceae)

Aph. – *Aphis* sp.

Apd. – *A. salicis*

(96) *Sisymbrium* sp. (Brassicaceae)

Aph. – *Myzus* sp.

Apd. – *Aphidius* sp.

Solanum species (Solanaceae)

(97) *Solanum nigrum* (Black Nightshade)

Aph. – *A. craccivora*

Apd. – *A. matricariae, T. (B.) indicus, L. gracilis*

Aph. – *Aphis fabae solanella* (Bean aphid)

Apd. – *T. (B.) indicus*, Pte. – *A. suspeusus* (hyperparasite)

Aph. – *M. persicae*

Apd. – *Toxares zakai*

(98) *Solanum tuberosum* (Potato)

Aph. – *A. craccivora*

Apd. – *D. rapae*.

Sonchus species (Asteraceae)

(99) *Sonchus asper* (Spiny-leaved Sow Thistle)

Aph. – *Uroleucon* sp.

Apd. – *P. volucre*

(100) *Sonchus oleraceus* (Milk-Thistle/ Common Sow-Thistle)

Aph. – *Nosonovia (Hypermyzus) lactucae(=Hypermyzus lactucae)*

Apd. – *P. volucre*

Aph. – *Uroleucon (Uromelan) sonchi*

Apd. – *D. rapae*

(101) *Sonchus* sp.

Aph. – *U. sonchi*

Apd. – *Aphidius* sp; Meg. – *D. cerpentri* (hyperparasite)

Sorghum species (Poaceae)

(102) *Sorghum halepense* (Johnson grass/Millet)

Aph. – *Sitobion avenae* (= *Macrosiphum avenae*)

Apd. – *A. ambiguus, L. confuscus*

(103) *Sorghum* sp.

Aph. – *Macrosiphum* sp.

Apd. – *A. ambiguus*

Aph. – *Melanaphis sacchari* (Sugar-cane Aphid)

Apd. – *L. confusus*

(104) *Spiraea* sp. (Rosaceae)

Aph. – *Aulacorthum* sp.

Apd. – *Ephedrus* (*E.*) sp.

Aph. – *Myzus cerasi* (Black Cherry aphid)

Apl. – *Aphelinus* nr. *fulvus*

(105) *Thuja orientalis* (Oriental Arbor vitae) (Cupressaceae) (=*Biota orientalis*)

Aph. – *Lachnus* sp.

Apd. – *P. (K.) hazartbalensis*

(106) *Trifolium* sp. (Clover) (Fabaceae)

Aph. – *A. pisum*

Apd. – *A. ervi*

(107) *Triticum aestivum* (=*vulgare*) (Poaceae)

Aph.–*Sitobion avenae*(=*Maicrosiphum avenae*)

Apd.–*Lysiphlebia mirzai*

Unidentified Aphid Species

Apd.–*Toxares shigai*

***Ulmus* Species (Ulmaceae)**

(108) *Ulmus villosa* (=*Ulmus laevigata*)

Aph. – *Eriosoma* nr. *ulmi*

Apd. – *A. lepelleyi*

(109) *Ulmus wallichiana* (The Himalayan Elm)

Aph. – *Schizonurrella indica.*

Apd. – *A. lepelleyi*

(110) *Urtica dioica* (Stinging Nettle) (Urticaceae)

Aph. – *Aphis* sp.

Apd. – *T. (B.) indicus*

Aph. – *Microlophium camosum*

Apd. – *A. ervi;* Cyn. – *Phaenoglyphis* sp., *matricariae, E. (E.) urticae*

Aph. – *Microlophium* sp.

Apd. – *T. zakai*; Cyn-*Alloxysta* sp. (hyperparasite)

(111) *Verbascum thapsus* (Cow's lungwort/ Common Mullein) (Scrophulariaceae)

Aph. – *A. nerii*

Apd. – *L. gracilis*

Apl. – *A. mali*

(112) *Veronica persicae* (Scrophulariaceae)

Aph. – *Aphis* sp.

Apd. – *A. gifuensis*

(113) *Viburnum grandiflorum* (Sheep berry, Stag Bush/Black Had) (=*Viburnum foetans*) (Caprifoliaceae)

Sc-i. – *Parthenolecanium corni* (Plum scale or European Fruit lecanium)

Apl. – *A. lingnanensis*

(114) *Vicia sativa* (Common Vetch, Spring Vetch) (Fabaceae)

Aph. – *Acyrthosiphon* sp.

Apd. – *A. ervi*

Aph. – *Aphis* sp.

Apd. – *A. gifuensis*

Aph. – *Therioaphis* sp.

Apd. – *T. (T.) complanatus*

(115) *Zinnia elegans* (Youth-and-old Age) (Asteraceae)

Aph. – *A. gossypii*

Apl. – *A. kashmiricus*; Apd. – *L. gracilis*

Unknown Host-Plant Species

Aph. – *Aphis craccivora*

Apl. – *A. (A.) abdominalis*

Aph. – *A. gossypii*

Apl. – *A. (A.) flavipes*

Aph. – *A. spiraecola*

Apl. – *A. (A.) flavipes*

Aph. – *Brevicoryne brassicae*

Apl. *A. (A.) asychis*

Aph. – *Eirosoma* nr *phenax*

Apd. – *A. lepelleyi*

Aph. – *Lipaphis erysimi*

Apl. – *A. (A.) asychis, A. (A.) flavipes*

Aph. – *M. sanbornii*

Apd. – *E. campestris, T. macrosiphophagus*

Aph. – *Myzus persicae*

Apl.— *A.(A.) asychis, A. (A.) flavipes*

Aph. – *Rhopalosiphum* sp.

Apl.—*A. (A.) flavipes*

Aph. – Unidentified aphid species

Apd. – *Adialytus arvicola, A. salicaphis*

Col. – unknown Wood-boring Coleopteran Species

Ich. – *Deuteroxorides indicus, Echthorus coracinus, Podoschistus scutellaris, Poemenia kashmirica*

Cor. – *Cletus bipunctatus*

Sce. – *Hadronotus* sp.

Lep. – Unknown Lepidopteran Species

Ich. – *Pimpla alternans, Scambus* (*Scambus*) *indicus* and *S.* (*S.*) *kashmiricus*

Pyr. – *Plutella xylostella* (Diamond black Moth)

Tac. – *Voria ruralis*

Ich. – *Horogenses* sp. and *Itoplectis* sp.

Sc-i. – *Lecanium* sp.

Enc. – *M. chalcostomus*

Sc-i. – *Ceroplastodes* sp. (Waxy Scale insect)

Apl. – *Coccophagus tschirchii*

Sc-i. – *Q. perniciosi*

Apl. – *Azotus kashmiricus, M. carnesi, P. chinensis*

Ten. – Unknown Sawfly Species

Ich. – *E. aithogaster, P. kashmirensis*

Unknown Host-plant & Insect Species

Bra. – *B.* (*B.*) *imitator, P. kashmirensis*; **Cha.**—*B. tachardia, C. ornatipes* and *E. kashmirensis*; **Con.** – *Ceria dimidiatipennis, Myopa nigriventris* and *M. testacea,* **Eul.** – *C. nitetis*; **Ich.** – *B. lactatorius, C. longipes, C. nursei, E. ruficornis, E. fornicator, G. kashmirensis I.fulvositigmata, P. montanes P. cellaris, P.kashmirensis, P. archica, P. albiventralis, S.nursei, X.* (*X.*) *authracinus, X.* (*X.*) *miniuus;* **Mym.** – *G. longicornis, G. munnarus, G. pahlgamensis, G. tarae, P. anantanogona,* **Mus.** – *Phaonia atronitens, P. thomsoni, P. flaromaculata, P. kashmirensis, Pogonomyia fumipennis, Hydrotaea armipes, H. dentipes, H.* (*Myiopsila*) *mediabunda angustifrons* and *Graphomyia maculata, Mydae* (*Myiospila*) *mediatabunda augustifrons.*

Key to the abbreviations (given in the catalogue):

Aph.– Aphid; **Apd.**– Aphidiid; **Apl.**–Aphelinid; **Bra.**–Braconid; **Cha.**–Chalcidid; **Col.**–Coleopteran; **Con.**–Conopid, **Cor.**–Coreid; **Cyn.**–Cynipidid; **Enc.**–Encyrtid; **Eul.**–Eulophid; **Eup.**–Eupelmid; **Eur.**–Eurytomid; **Hym.**–Hymenopteran; **Ich.**–Ichneumonid; **Lep.**–Lepidopteran; **Lip.**–Liparid; **Lym.**–Lymantriid; **Meg.**–Megaspilid; **Mus.**–Muscidid; **Mym.**–Mymarid; **Pte.**–Pteromalid; **Sar.**–Sarcophagid; **Sce.**–Scelionid; **Sc-i.**–Scale insect; **Sco.**–Scotylid; **Tac.**–Tachnid; **Ten.**–Tenthredinid; **Tor.**–Tortricid; **Torm.**–Torymid and **Tri.**–Trichogrammatid.

BIBLIOGRAPHY

Abrol, D.P. & Kotwal, D.R. (1996). Occurrecne of *Varroa jacobsoni* Oudemans in Jammu, India on honey bees. *Insect Env.,* **2**(2):48.

Abrol, D.P. & Putatunda, B.N. (1995). Discovery of ectoparasite mite *Tropilaelaps koenigenus.* Dilfinado Baker and Baker (Acari: Laelapidae) on *Apis dorsata, A. mellifera* and *A. cerana* F. in Jammu & Kashmir, India. *Curr. Sci.*, **68**(1):90.

Agarwal, B.L. (1951). Relapsing fever in Kashmir. *Indian Med. Gaz.,* **86**:446-53.

Ahmad, Bashir, Bhatti, Gurjeet, Thokar, M.A. and Malla, Nancy (2002). Human Toxocariasis and ascariasis: concomitant parasitism in Srinagar, Kashmir, India. *Indian J. Pathol. Microbiol.*, **45**(31):315-318

Ahmad, D. & Dar, H.U. 1975 (1978). Biological Control of potentially serious orchard pest in Kashmir and feasibility of integrated control. *J. Sci.Univ. Kashmir.,* **3**(2):93-98.

Ahmad, Fayaz (1993). *Trematode and Cestode Parasites of Pisces and Aquatic aves of Kashmir, with Histology and Histopathology of Genus Adenoscolex* Fotedar, 1958, Vol. I, 430 p. Vol. II, 488 pp. + I-XCVII plates.

Ahmad, Fayaz (2000). Avian Trematode Parasites of Kashmir. Part VI. Genus *Paryphostomum* Dietz, 1909. *J. Parasitic Diseases,* **24**(2):155-158.

Ahmad, Fayaz & Chishti, M.Z. (1995). Avian Trematode Parasites in Kashmir. Part.II. Genus *Notocotylus* Diesing, 1839. *Oriental Sci.,* **1**(1-2):59-65.

Ahmad, Fayaz & Chishti, M.Z. (1995a). Avian Trematode Parasites of Kashmir Part I: The genus *Paramonostomum* Luhe, 1909, with description of a new species *P. zahoorensis (Proc.* 12th *Natl. Congr. Parasitol.),* held at Panaji, Goa, from 23-25th Jan., 1995). *J. Parasitic Diseases,* **19** (81-105).

Ahmad, Fayaz & Chishti, M.Z. (1997). Avian Trematode Parasites of Kashmir. Part III. Genus *Echinoparyphium* Dietz. 1909. *Oriental Sci.,* **2**(1):41-48.

Ahmad, Fayaz & Chishti, M.Z. (1997*a*). Avian Trematode Parasites of Kashmir. Part IV. *Echinostoma revolutum*, (Forelich, 1802) Looss, 1899. *Oriental Sci.,* **2**(2):49-59.

Ahmad, Fayaz & Chishti, M.Z. (2000). Avian Trematode Parasites of Kashmir. Part V. Genus *Euparyphium Dietz* 1909). *J. Parasitic Diseases,* **24**(2):219-220.

Ahmad, Fayaz & Chishti, M.Z. (2000a). Fish Trematode Parasites of Kashmir. PartII. Genus *Clinostomum* Leidy, 1856 (Digenea: Clinostomatidae). *Oriental Sci.*, **5**(1):13-22.

Ahmad, Fayaz & Chishti, M.Z. 1999 (2001). Fish Trematode Parasites of Kashmir. Part-I. Genus *Diplozoon* Nordmann, 1832 (Monogenea, Poly ophisthocotylea) *Orient. Sci.*, **4**(2):79-91.

Ahmad, Fayaz & Dhar, R.L. (1996). Avian Trematode Parasites of Kashmir. Part IV. Genus *Echinostoma* Rudolphi, 1809. *J. Parasitic Diseases,* **20**(1):113.

Ahmad, Ishtiaq, Ahmad, Farooq, Mir, A. H., Wani, N. A., Rifat & Mir. N. A. (2001). Solitary hydatid sigmoid mesocolon. *JK-Practioner,* **9**(1): 49-50.

Ahmad, Muzzafar (1978). *Incidence of Helminthic* and *protozoan infestation in gut in rural population of Kashmir*. M.D. Medicine Thesis. Department of Medicine Government Medical College. The University of Kashmir, Srinagar. p.111.

Akhtar, Yasmin, Ahmad, W. & Jairajpuri, S. (1996). Two New Species of Dorylaimida (Nematoda) from Kashmir Valley, India. *Rec. Zool. Surv. India,* **94**(1):87-92.

*Ali, Neelofer (1999). *Parasitism in Aquatic vertebrates of Anchar lake.* Ph. D. Thesis. P. G. Deptt. of Zool., Faculty of Sci., Univ. of Kmr., Sgr., Kmr. (J & K).

Ambasta, S.P. *et al.* (eds) (1986). *The Useful Plants of India.* Published by Publication and Information Directorate of Scientific & Industrial Research, New Delhi (India).

Anderson, R.C. (2000). *Nematode Parasites of Vertebrates: Their Development and Transmission* (2nd Edition). CBI Publishing, CAB International Wallingford, Oxan, U.K. 650 p.

Anonymous (1969). *Report of Work carried out during 1968.* Commonwealth Institute of Biological Control, Commw. Agri. Bureaux, London, England, 100 p.

Anonymous (1970). *Report of Work carried out during 1969.* Commonwealth Institute of Biological Control, Commw. Agri. Burx., London, England, 183 p.

Ara, Jahan, Khan, A. R. & Ahmad, Fayaz (2000). First record of Pseudophyllidean cestode, *Bothriocephalus* (Rudolphi, 1808) from fishes of Kashmir. *Oriental Sci.,* **5**(1):23-26.

Arora, R.R., Biswas, M., Mathur, K.K. Verghese, T. & Das, A. (1976). Prevalence of intestinal parasites in rural community in Jammu and Kashmir. *J. Communicable Diseases,* **8**(4):302-315.

Baer, W.S. (1931). The treatment of chronic oesteomyletis with maggots (larva of blowfly). *J. Bone & Joint.,* **13**: 438-475.

*Bakshi, Shafqat (1999). *Fish Parasites of some Lakes of Kashmir, with an analysis of seasonality of incidence with regard to different ecological factors.* Ph. D. Thesis, Faculty of Sci., P. G. Deptt. of Zool., The Univ. Kmr., Sgr., Kmr. (J&K).

Bali, H.S. (1971). On the occurrence of *Calicophoron caliciphorum* (Fishoeder, 1901) Nasmark, 1937 in sheep in J&K State. *Indian Vet. J.,***49**(4):442-43.

Bali, H.S. (1972). *Helminth Parasites of sheep in Jammu & Kashmir.* Ph. D. Thesis. P. G. Deptt. of Zool., Faculty of Sci., Univ. of Kmr., Sgr., Kmr. (J&K).

Bali, H.S. (1972a). On the morphology of a new trichostrongylid, *Nematodirus fotedari* n. sp. of sheep *(Ovis aries)* from Kashmir. *J. Res. Panjab Agri. Univ.,* **9**:195-198.

Bali, H.S. (1972b). Study on a new trematode belonging to the genus *Paramphistomum* Fishoeder, 1901 from *Ovis aries* in J&K State. *Indian J. Anim. Sci.,* **42**(3):231.

Bali, H.S. & Fotedar, D.N. (1972a). On the morphology of a Paramphistome in sheep. *Ceylonocotyle scoliocoelium* from Jammu & Kashmir *J. Res. Punjab. Agri. Univ.,* **9**(1):199-200.

Bali, H.S. & Fotedar, D.N. (1973). On a new Amiphistome, *Olveria thapari* n.sp. from rumen. of cross-bred sheep from Kashmir. *Indian J. Helminthol.,* **24**(1):36-39.

Bali, H.S. & Fotedar, D.N. (1974). Effect of experimental infection of some bursate nematodes in sheep in Kashmir. *J. Res. Panjab. Agri. Univ.,* **11**(1): 114-119.

Bali, H.S. & Fotedar, D.N. (1974a). Studies on Juvenile stages of some helminth parasites of Kashmir sheep. *J. Res. Punjab Agric. Univ.,* **11**(8):329-343.

Bali, H.S. & Fotedar, D.N. (1976). Studies on the pathology of some helminth-borne disease in sheep of Kashmir. *J. Anim. Health Prod.* P. A.U., **4**(1-2):9-14.

*Bambroo, Neena (1970). *Helminth parasites of domestic animals.* Ph. D. Thesis. P. G. Deptt. of Zool., Faculty of Sci., Univ. of Kmr., Sgr., Kmr. (J&K).

Bambroo, Neena (1970a). On two species of *Gastrothylax* Poinier, 1883 from sheep in Kashmir. *Kashmir Sci.,* **7**(1-2):147-155.

Bambroo, N. (1971). On a new bursate nematode from sheep in Kashmir. *Indian J. Helminthol.,* **23**(2):86-94.

Bansal, G.C., Sharma, R.L. & Dhar, D. N. (1982). Lungworms—Threat to sheep and goats. *Farmers Journal,* **1**:42-44.

Baqui, A. & Shah, Farooz, A. (2002). Alteration in haematological parameter in nematodiasis. *Oriental Sci.,* **7**(1):25-34.

Barraud, P.J. (1924). A new mosquito from Kashmir. *Indian J. Med. Res.,* **11**: 967-968.

Barraud, P.J. (1924 *a*). A new mosquito from Kashmir and the North-West Frontier Province. *Indian J.Med. Res.,* **12**: 73-74.

Basir, M. A. (1941). Two new nematodes from an aquatic beetle. *Proc. Ind. Acad. Sci.,* **13**: Sec. B., 163-167.

Bhadwal, M.S. & Raina, A.K. 1994: Thelaziasis in a buffalo. *Indian J. Vet. Med.,* **14**(2):94.

Bhagat, K.C. (2000). Natural parasitism of wheat aphid, *Macrosipham (Sitobion) avenae* Fab. by *Lysiphlebia mirzai.* Shuja-uddin (Hymenoptera: Braconidae). *Insect Envir.,* **6**(1):9.

*Bhagat, R.C. (1980). *Parasitoid – complex of aphids of Kashmir, with special reference to Aphidiidae (Hymenoptera).* Ph. D. Thesis. P. G. Deptt. of Zool., Faculty of Sci., Univ. of Kmr., Sgr., Kmr. (J & K).

Bhagat, R.C. (1981a). A new species of *Monoctonus* (Hymenoptera: Aphidiidae) from Kashmir, India. *Curr. Sci.,* **56**(24):1087-1088.

Bhagat, R.C. (1981b). *Areopraon lepelleyi* (Hymenoptera: Aphidiidae): a new record for India. *Geobios,* **8**(4):177-178.

Bhagat, R.C. (1981c). Some new aphid parasitoids. (Hymenoptera: Aphidiidae: from Kashmir, India. *Orient. Insects,* **15**(2):195-200.

Bhagat, R. C. (1982). Aphid-galls and their parasitoids from Kashmir. *Entomon,* **7**(1):103-105.

Bhagat, R. C. (1982a). On a new aphid parasitoid (Hymenoptera: Aphidiidae) from Kashmir. *Indian J. Zool.,* **10**(1):19-22.

Bhagat, R. C. (1982b). On two new aphid parasitoids of Genus *Trioxys* (Aphidiidae: Hymenoptera) from Kashmir, India. *Entomon,* **7**(3):321 -324.

Bhagat, R. C. (1982c). New aphid parasitoid (Hymenoptera: Aphidiidae) from Kashmir, India. *Oriental. Insts.,* **16**(1):113-117.

Bhagat, R. C. (1983). Records and host range of hyperparasitoids (Insects: Hymenoptera) of aphids (Homoptera: Aphidiidae) from Kashmir, India. *Sci. & Cult.,* **49**(3):150-151.

Bhagat, R. C. (1984). New records and hosts of aphids parasitoids (Hymenptera: Aphidiidae) from Kashmir, India. *J. Bombay Hist. Soc.,* **81**(1):93-98.

Bhagat, R. C. (1985). Host-aphid complex of Aphelinid parasites (Hymenoptera: Insecta) in Kashmir Valley, India. *Geobios New Reports,* **4**:176-179.

Bhagat, R. C. (1985a). On Aphids and their parasitoids. (Aphidiidae: Hymenoptera), infesting forest-trees in Kashmir Valley, India. *Indian Forester,* **111**(6):467-473.

Bhagat, R. C. (1986). Additional records and host aphids and aphidiid parasitoids from Kashmir Valley, India. *Geobios new Repts.,* **5**:161-162.

Bhagat, R. C. (1986a). Aphid pests of fruit trees and their natural enemies in Kashmir Valley, India. *J. Agri. Sciences,* **56**(7):532 -536.

Bhagat, R. C. (1986b). On aphid pests and their aphidiid parasitoids of agricultural importance. *Indian Agriculturist,* **30**(3):229-235.

Bhagat, R. C. (1987). Field observations on hyperparasitoids of aphid pests, infesting *Rosa* spp. in Kashmir Valley. *Indian. J. Biol. Control,* **1**(2):104 -105.

Bhagat, R. C. (1987a). On aphid damaging medicinal plants and their parasites from Kashmir Valley, India. *Geobios New Repts.,* **6**:11-14.

Bhagat, R. C. & Ahmad, Nissar (1995). Aphidiid parasitoids (Hymenoptera) of aphids (Homoptera) of Jammu – New record, host range and biological notes. *J. Aphidology,* **5**(1&2):90-96.

Bhagat, R. C., Farhan, N. & Ramzan, A. (1992). Natural enemies of scale insects in Kashmir Valley, India. *J. Biol. Control,* **6**(1):38-39.

Bhalerao, G. D. (1973). Studies on the Helminths of India, Trematoda IV. *J. Helminthol.,* **15**: 97-124.

Bhat, H. R. & Kulkarni, S. M. (1971). A report on the occurrence of *Culex (Neoculax) hortensis* Fiscalbi, 1889, in Ladakh. *Oriental Insects,* **5**(4): 583-584.

Bhat, H. R. & Kulkarni, S. M. (1983). Records of mosquitoes collected from Jammu & Kashmir, with ecological notes. *Recd. Zool. Surv. India.* **81**:(1-2): 192-203.

Bhat, O. K., & Kaul, V. (1994a). Influence of soil temperature on the population of root knot nematode, *Meloidogyne icognita* in a tomato field. *Annl. Plant Prot.,* **2**(1):28-29.

Bhat, O. K., Kaul, V. & Bhagat, K. C. (1994). Incidence of pests associated with the rhizosphera of tomato in Jammu. *Annls. Plant Protec. Soc.,* **2**(2).

Bhat, Shama (1979). The genus *Phaenocarpa* Foerster (Hymenoptera: Braconidae, Alysiinae). *Oriental Inst.,* **13**(1-2):1-27.

Bhatia, M. L. (1930). Surune nouvella hirudinec thynchobdellida, *Glossiphonia cruciata* n. sp. provenant du vivier a truiteid Achhabal Kashmir. *Annls Parasit. Kum. Camp.,* **8**:244-348.

Bhatia, M. L. (1934). Nouvella sangue rhynchobdellida *Glossiphonia lobata* n. sp. de Testablispiment de pisciculture Achabal. *Annls. Parasit. Kmr. Camp.,* **12**:121-129.

Bhatia, M. L. (1939). On some leeches from Kashmir. *Bull. Punjab Univ. Zool.,* G **2**:1-17.

Boucek, Z., Subba Rao, B. R. & Farooqi, S. I. (1978). A preliminary review of Pteromalidae (Hymenoptera) of India and adjacent countries. *Oriental Insects,* **12**(4):433-468.

Boxshall, G.A. (1976). A new genus and two new species of copepod parasitic on fresh water fishes. *Bull. Br. Mus. Nat. Hist.* (Zool.) **30**(6):209-215.

Brunnetti, E. (1920). *The Fauna of British India, including Ceylon and Burma. Part-III. Pipunculidae, Syrphidae, Conopidae & Cestridae.* Taylor & Francis, London, England.

Buhroo, A. A., Masoodi, M. A. & Chishti, M. A. (2001). Description of two new species of chalcidoidae (Hymeroptera) from fruit orchard of Kashmir. *SKAUST J. Res.,* **3**(1):14-20.

Buhroo, A. A., Chishti, M. Z. & Masoodi, M. A. (2002). Biocontrol agents of shot-hole Borer, *Scolytus nitidus* Schedl (Coleoptera: Scolytidae), infesting apple orchards. *Indian J. Plant Protection,* **30**(1):45-48.

Cameron, P. (1906). On the Tenthedinidae parasitic hymonptera in Baluchistan by Major C. G. Nurse. *Bombay Nat. Hist. Soc.,* **17**(2):274-280.

Cameron, P. (1906a). On the Tenthedinidae and parasitic hymenoptera collected by Major C. G. Nurse in Kashmir, *Bombay Nat. Hist. Soc.,* **17**(2):289-292.

Chakravorti, G. K. (1942). A new nematode *Camallanus salmonae* from Kashmir. *Curr. Sci.,* **11**: 441-442.

Chandler, A.C. & Read, C.P. (1961). *Introduction to Parasitology, with special reference to the parasites of man.* John Wiley and Sons Inc., Indian Reprint Edition 1970, Wiley Eastern Private Ltd., New Delhi, 822 p.

Chandniwala, K. M. (1996). *Plant Nematodes – Recent Advances in Plant Diseases.* Anmol Publication Pvt. Ltd. New Delhi, India.

Chandra, M. (1982). A checklist of leeches of India. *Recd. Zool. Surv. India.,* **80**:265-270.

Cheng, C.T. (1973). *General Parasitology.* Academic Press Inc. U.S.A. p. 917.

Chishti, M. Z. (1964). *Cestodes and Nematode Parasites of some Aves of Kashmir.* Ph. D. Thesis, Faculty of Science, P. G. Deptt. of Zool., The Univ Kmr., Sgr., Kmr. (J&K), 319 p.

Chishti, M.Z. (1973). On a new species of *Choanotaenia* Railliet.,1896 from *Acridotheres tristis* in Kashmir. J. Sci. Kmr., **1**(1-2):51-54

Chishti, M. Z. 1975 (1978). A new record of *Acuaria skrjabini* Ozerskeya, 1926 from *Passer domesticus* in Kashmir. *J. Sci. Univ. Kashmir,* **1**(1-2): 109-116.

Chishti, M.Z. (1980). *Dilepsis fotedari* n. sp. (Dilepididae Furhmann, 1907: Cestoda) from *Anas platyrhynchos* in Kashmir. *Indian J. Helminthol,* **36**(1):1-3.

Chishti, M. Z. (1981). On a new record of *Syngamus trachea* (Montagu, 1811) Chapin 1925 (Syngamidae, Leiper, 1912): Nematoda) from *Corvus splendens* in Kashmir. *Kashmir Univ. Res. J.,* **1**(1):53-55.

Chishti, M.Z. (1982). *Pseudoschistotaenia pindchii* sp. nov. (Acoela: cestoda) from *Podiceps ruficollis capensis* in Kashmir. *Osnovy Cestolognii,* **11**:69.

Chishti, M. Z. & Abdullah, R. (1986). On *Rusguniella (Rusgunioides) turdi* sp. nov. (Acuarioidea: Nematode) from *Turdus unicolor* in Kashmir. *Indian J. Helminthol.,* **38**(1): 79-82, (1986/87).

Chishti, M.Z. & Ahmad, F. (1992). A new record of *Pseudoechinostomum indicum* Mehra, 1952 from *Anas platyrhynchos* in Kashmir. *Indian J. Helmintol.,* **44**(2):101-104.

Chishti, M.Z. & Ahmad, F. 1991 (1992a). On a new species of *Echinostoma kashmirensis* sp. n. (Echinostomatidae Poche, 1926) from *Anas platyrhynchos domesticus* in Kashmir. *Indian J. Helminthol.,* **43**(2):153-155.

Chishti, M.Z., Ahmad, Fayaz & Bakshi, Shafqat (2002). Comparative Fish Parasite Diversity in two fresh water Lakes in Kashmir, in relation to ecological factors in different seasons. *Proc. 10th Intl. Congr. Parasitol.* ICOPAX; Vancouver (Canada), August 4-9, 2002, p. 379-386.

Chishti, M. Z. & Bakshi, S. (1990). On *Rhabdochona guptai* n. sp. from the intestine of fish *Cyprinus carpio specularis,* fish from Hokersar Wetland in Kashmir. *Indian J. Parasitol.,* **14**(1): 45-49.

Chishti., M.Z. & Bakshi, Shafaqat (1992). New species of the genus *Pomphorhynchus* Monticelli, 1905 from *Cyprinus carpio communis,* a fresh water fish in Kashmir. pp. 95-102, *In:* Yousuf, A.R., Raina, M.K & Qadri, M.Y. (Eds.), *Curr. Trends in Fish and Fishery Biology and Aquatic Ecology (Proc. 3rd. Natl Ser. Fish* and their Env. field at P.G. Deptt. of Zool. The Univ. of Kashmir, Srinagar, Kashmir, from 19-22. Sept., 1989.

Chishti, M. Z. & Khan, A. R. 1982 (1984). *Mayhewia kavini* sp. nov. (Hymenolepidae: Raillet et Henry 1909*)* from *Corvus monedula* in Kashmir. *Indian J. Helminthol.,* **34**(2):107-111.

Chishti, M.Z. & Mir, A.A. 1989 (1990). On a new species of *Echinochasmus (Echinochasmus)* Dietz, 1909 (Echinostomatidae, Echinoschaminae Odhunner, 1911) from little grebe *Podiceps ruficollis capensis* in Kashmir, *Indian J. Helminthol.,* **41**(2):141-145.

Chishti, M.Z., Mir, A.A. & Asmat Rasool, 1986 (1987). *Choanotaenia micracantha* sp.nov. (Dilepoidea: Cestoda) from *Corvus monedula* in Kashmir. *Indian J. Helminthol.,* **38**(2):107-111.

Chishti, M. Z. & Naqash, A. R. (1982). *Varroa jacobsonii*— a serious menance to *Apis cerana* in Kashmir, *Kashmir Univ. Res. J.,* **2**:43-45.

Chishti, M.Z. & Peerzada, M.Y., 1998. Host and seasonal occurrence of Acanthoccephala in fishes of Wular Lake. *Oriental Sci.,***3**(1):31-38.

Chishti, M. Z., Tak, Hidayat-ullah & Ahangar, A. G. (2001). Epidemiology of hydatid disease in Human Population of Kashmir Valley. *Oriental Sci.,* **6**(1):1-8.

Chishti, M.Z. &. Tanveer, S. (1992). On a new species of trematode genus *Echinoparyphium Dietz* 1909 from domestic fowl in Kashmir. *Indian J. Parasitol.,***16**(2):155-157.

Chishti, M.Z. & Tanveer, Syed. (1993). On *Paryphostomum srivastavi* n.sp. from *Gallus gallus domesticus. J. Indian J. Helminthol.,* **45**(42):152-54.

Chitwood, B.G. (1950). An outline classification of nematoda. pp.15-25, *In:* Chitwood, B.G. & Chitwood, M.B. (Eds.). *Introduction to Nematology.* University of Park Press, Baltimore, Maryland. U.S.A.

Christopher, S. R. (1933). *The Fauna of British India, Diptera,* Vol. IV, *Family Culicidae, Tribe Anophelini.* Tylor and Francis, London, England.

D 'Hondt, J.L. (1997). Qu'est-il advanu des anciens-Aschelminthes. *Bulletin de la Societe Biologique France*, **122:** 355-364.

Das, S. M. & Sharma, B. D. (1970). A new record of *Ctenocephalides felis orientis* (Jordon, 1925), (Siphonaptera; Pulicidae) from goats in Kashmir. *Kashmir Sci.,* **7**: 75-78.

Das, S.M. & Sharma, B.D. (1970a). A survey of ticks of Kashmir. *Kashmir Sci.,***7**:111-118.

Das, S. M. & Sharma, B. D. (1970b). On *Haematopinus eurysternus* Nitzsch (1818) (Siphunculata: Haematopinidae) from Kashmir. *Kashmir Sci.,* **7**(1-2): 79-81.

Das, S.M. & Shaw, A.A. (1968). Studies on Parasitic Protozoan of Domestic animals of Kashmir. I. On the occurrence of *Babesia argentina* Ligenieres (Protozoa, Haemosporida) in imported Australian Jersey Cows. *Kashmir Sci.,* **5**(1-2):238-292.

Datta, M. (1983). A review of the Simuliidae (Dipter) from Oriental Region. *Oriental Insects,* **17**:215-267.

Datta, M. 1985 (1986). Notes on blackflies (Diptera: Simuliidae) from Kashmir, India. *Bull. Zool. Surv. India,* **7**(2-3): 231-236.

Datta, M. & Chakraborti, M. (1983). On a collection of flowerflies (Diptera: Syrphidae), with new records from Jammu and Kashmir. *Rec. Zool. Surv. India,* **81**: 237-253.

Datta, M.N. (1936). Scientific results of the Yale North India Expedition. Biological Rept. No. 20. Helminth parasites of fishes from North-India, with special reference to Acanthocephalans. *Rec. Indian Mus.,* **38**:211-229.

Datta, M. N. (1937). Acanthocephala from India. Description of male of *Eosentis rigidus* van Cleave, occurring in the intestine of *Schizothorax zarudnyi* from Seistan *Rec. Ind. Mus.,* **39**:303-304.

Dhanda, V. & Bhat, H.R. (1971). *Haemaphysalis (Herpetobia) sundrai* Sharif (Ixodoidea: Ixodidae), a tick parasitizing sheep in the Western Himalaya, redescription of female, description of male and ecological observation. *J. Parasitol.,* **57**(3):646-650.

Dhanda, V. & Kulkarni, S.M. (1969). Immature stages of *Haemaphysalis cornupunctata* Hoogstraal & Varma, 1962 (Acarina: Ixodiadae), with new host and locality records and notes on its ecology. *Oriental Insects.,* **3**(1):15-21.

Dhanda, V., Kulkarni, S.M. & Kaul, H.N. (1979). Record of *Nosomma monstrosum* (Acarina: Ixodidae) from Jammu & Kashmir. *Acarology Newsletter,* No. 8:3-4.

Dhanda, V., Kulkarni, S.M. & Pratt, P. (1971). *Dermacentor raskemensis* (Ixodoidea: Ixodidae), redescription and notes on ecology. *J. Parasitol*., **51**(6):1324-29.

Dhar, D.N. (1980). Recent studies on lungworm infection of sheep and goats in India. *Indian J. Parasitol.,* **3**:127-128.

Dhar, D.N. (1982). Recent studies on *Dictycaulus filaria* and other lungworm of sheep and goats in India. *J. Nuclear Agric. & Biol.,***11**:1-5.

Dhar, D. N. (1985). Bovine and ovine parasitic bronchitis, with particular reference to the work done in India. pp. 53-88, *In.* Sen, A. B., Katiyar, J. C. & Guru, P. Y. (Eds.), *Prespective in Parasitology.* Print House (India), Lucknow, India.

Dhar, D.N. & Sharma, R.L. (1978). *Dictyocaulus filaria* in lambs. The effect of varying single infection on subsequent production. *Vet. Parasitol.,* **4**:221-229.

Dhar, D.N. & Sharma, R.L. (1978a). Studies on the comparative susceptibility of sheep and goats to infection with *Dictyocaulus filaria. Ind. J. Anim. Sci.,* **48**:29-31.

Dhar, D.N. & Sharma, R.L. (1979). A note on the prevalence of lungworm infections in sheep and goats in Tehsil Kargil of district Ladakh (Jammu & Kashmir). *Indian J. Anin. Sci.,* **49**:585-588.

Dhar, D. N., Sharma, R. L. & Bansal, G. C. (1979). Occurrence of *Mullerius* sp. in sheep in India. *Indian J. Parasitol.,* **3**:187.

Dhar, D.N., Sharma, R.L. & Bansal, G.C. (1982). Gastro-intestinal nematodes in sheep in Kashmir. *Vet. Parasitol.,* **10**:271-277.

Dhar, D. N., Sharma, R. L. & Bansal, G. C. (1983). Ascarid infection in sheep in Kashmir Valley. *Indian J. Parasitol.,* **7**(1):97-99.

Dhar, D.N., Tewari, H.C. & Rao, Y.V. B.G. (1972). Prevalence of lungworm infections in sheep in Kashmir. *Indian. J. Animal Sci.,* **42**:112-114.

*Dhar, R.L. (1973). *Studies on parasites of fishes of Jammu & Kashmir.* Ph.D. Thesis, P.G. Deptt. of Zool., Univ. Kmr., Srg., Kmr. (J&K).

Dhar, R. L., 1975 (1978). *Astiotrema fotedari* n. sp. (Trematoda: Plagiorchidae Luhe, 1901) from the intestine of *Laleo dero* (Gemilton). *J. Sci. Univ. Kashmir,* **3**(12):99-104.

Dhar, R.L. & Ahmad, F. (1992). A new record of *Echinoparyphium clerci* Skrybin from *Anas platyrhynchos* in Kashmir. *Indian J.Helm.,* **44**(2):93-96.

Dhar, R.L. & Ahmad, F. (1992a). First record of *Cotylurus cornutus* Rudolphi, 1887 Szidal, 1928, from *Anas platyrhynchos domesticus* in Kashmir. *Indian J. Helminthol.,* **44**(2):105-108.

Dhar, R.L. & Ahmad, F., 1991(1992). *Pseudoechinostomum srinagarensis* sp. nov. (Trematoda:Echinostomatidae Looss, 1902) from a domestic duck *Anas platyrhynchos domesticus* from Kashmir. *Indian J. Helmintol.,* **43**(2):147-49.

Dhar, R. L. & Anwari, M. D. (1982). On new host record of *Rhaldochona* Raillict, 1916. (Rhaldochonidae Skrjabin, 1946, Nematode) *Kashmir Univ. Res. J.,* **2**:39-40.

Dhar, R.L. & Fotedar, D.N., (1979). On *Procammalanus(Monospiculus) kashmirensis* sp. nov from fresh water fish, *Wllago attu* from Jammu, India. *Indian J. Helminthol.*, **40** (2):128-134.

Dhar, R.L. & Fotedar, D.N. 1979 (1980). Some protocephalid cestodes in fresh water fishes of Jammu and Kashmir. *Indian J. Helminthol.,* **31** (2):111-127.

Dhar, R.L. & Kharoo, V.K. (1985). A new species of Acanthocephala, *Neoechinorhynchus glyptosternumi* n .sp. from Intestine of Kashmir Fish, *Glyptosternum* sp. *Indian. J. Helminthol.*, **36**(1):36-39.

Dhar, R. L. & Kharoo, V. K., (1985a). On a new trematode *Allocreadium fotedari* sp. nov. of the family Allocreadiidae Stossich, 1903, from the intestine of *Schizothorax niger,* a fresh water fish of Kashmir, Inida. *Indian J. Helminthol.,* **36**(1):32-35.

Dhar, R.L. & Kharoo, V.K. 1986 (1987). Studies on trematode parasites of fishes, genus *Clinostomum* Keidy 1856 from fresh water fishes of Kashmir. *Indian. J. Helminthol,* **38**(1):74-78.

Dhar, R.L. & Kharoo, V.K. (1991). Trematodes of aquatic and migratory birds in Kashmir. *Trudy Gel mintologicheskoi Laboratorii.* No. **38,** 28-32.

Dhar, R.L. & Majidah (1983). *Gangesia (Gangesia) fotedari* sp. nov. (Protocephalidae; Cestoda) from fresh water fish in Wular Lake, Kashmir. *Indian. J. Helminthol.*, **35**(1):73-77.

Dhar, R.L. & Majidah, Rasheed (1987). Fish Parasitization of Helminths in Wular Lake, Kashmir. India. *Indian J. Helminthol.,* **34**(2):143-152.

Dhar, R. L. & Peerzada, M. Y. (1992). Seasonal variation in the occurrence and maturation of *Adenoscolex oreini* Fotedar, 1958 (Cestoda: Caryophyllaeidae) infecting some cyprinid fishes of Wular Lake (Kashmir), pp.103-107. *In*: Yousuf *et al.* (eds.) *Current trends in Fish and Fishery Biology and Aquatic Ecology* (*Proc. 3rd. Natl. Semr.* on *Fish and their Env*.), held at Srinagar, Kmr. on 19-22 September, 1989, at P. G. Deptt. of Zool., Univ. of Kmr., Sgr., Kmr. (J&K).

Dhar, R.L. & Peerzada, M.Y. 1993(1994). Seasonal variations in helminth population of common snow trout, *Schizothorax niger* Heckel of Wular Lake, Kashmir. *Indian. J. Helminthol.*, **45**(41/2):86-91.

Dhar, R.L. & Raina, M.K. (1983). On a new cestode, *Protocephalus kashmirensis* sp. nov. from a fresh water fish, *Nemachilus kashmirensis (Noemacheilus kashmirensis)* (Hora) from Kashmir, India. *Indian J. Helminthol.,* **35**(1):30-34.

Dharmadhikari, P. R. & Ramaseshia, G. (1970). Recent records of aphidiids (Hym; Aphididae) in India. *Tec. Bull. Commonw. Inst. Biol. Control,* **13**:(83-89.

Diller, E. (1984). Orientalische Taxa der Gattuug. *Promethes* Forester, (1989) (Hymenoptera, Chneumonidae, Diplzontinae). *Entomofauna*, **5**(6):69-80.

*Dilshada (1997). *Studies on the microbial and helminth infections of fishes in some lakes of Kashmir and ecological characteristics of water infecting them.* Ph. D. Thesis, P.G. Deptt. Zool. Univ. of Kmr., Srg Kmr., (J&K).

Dutta, R.N., Mukerjee, D.G. & Bhattacharjee, S.N. (1963). Tick-borne relapsing fever. *Armed Forces Med. J. India.*, **19**:125-33.

Eastop, V. F. & Hille Ris Lambers (1976). *Survey of World Aphids.* Dr. W. Junk. b. v. Publishers. The Hague, Netherland.

Edwards, F. W. (1927). Two new *Simulium* from Kashmir. *Bull. Ent. Res.,* **18**: 169-170.

Evans, G. O. (1992). *Principles of Acarology.* CAB International, Willingford, Oxon, U.K.

Faust, E.C. (1927). Studies on Asiatic Holostomes class Trematoda. *Rec. Indian Mus.,* 215-218.

Faust, E. C., Russell, P. F. & Jung, R. C. (1974). *Craig and Fausts clinical parasitology* (8th Edition). Lea & Febiger Philadelphia, U.S.A., 890 pp + 340 illust, 8 col. Plts.

Fazili, A., Wani, N. A., Tahir, M. S., Khan, Saleem & Mir, A. R. (2002). Hydatidosis – rare presentation. *JK Practioner,* **9**(4): 252-253.

Ferris, G. F. (1951). The sucking lice. *Mem. Pacif. Coast. Ent. Soc.,* **1**: 1-320.

Fomda, B. A., Sofi, B.A., Thokar, M. A. & Kakru, D. K. (2002). Human hydatid disease in Kashmir North India – A hospital based study. *J. Parasitic Diseases,* **26**(1): 34-37.

Fotedar, D. N. (1958). On a new caryophyllaeid cestode, *Adenoscolex oreini* gen. et sp. nov. from fresh water fish in Kashmir and note on the related genera. *J. Helm.,* **32**:1-16.

Fotedar, D.N. (1959). On a new species of the genus *Ganeo* Klein and some notes on the genus. *J. Helminthol.,* **33**(2/3):151-160.

Fotedar, D. N. (1960). On a new species of *Oxysomatium* Railliet and Henry, 1911 and some notes on the genera *Oxysomatium* and *Aplectana., J. Helminthol.,* **34**:141-150.

Fotedar, D. N. (1964). A new species of the nematode genus, *Pseudonymus* Diesing 1857 from an aquatic beetle in Kashmir, *Kashmir Sci.,* **1**(1-2): 73-75.

Fotedar, D.N. (1964a). On a new species of the trematode genus *Typhlocoelum* Stossich, 1902 from mallard duck in Kashmir and a review of previous work on the genus. *Kashmir Sci.,* **1**(1&2):44-53.

Fotedar, D.N. (1965). On the larval stages of a Brachylaemid trematode in the land snail, *Buliminus dextrosiniater,* in Kashmir. *Kashmir Sci*., **2**:101-113.

Fotedar, D. N. (1965a). *Rhabdias bufonis* (Schrank, 1788). Stiles and Hassall,1905 from the lungs of *Bufo viridis* in Kashmir and notes on some amphibian species of the genus. *Kashmir Sci.,* **2**(1-2): 20-26.

Fotedar, D. N. (1966). New Nematotaeniid Cestode from *Bufo viridis* in Kashmir. *Kashmir Sci.,* **3**(1-2):77-82.

Fotedar, D.N. (1967). On the morphology of *Prostocus himalayai* Pande, 1937 and synonymy of two species of the genus. *Kashmir Sci.,***4**(1-2):227-234.

Fotedar, D.N. (1968). New species of *Neoechinorhynchus* Hamann, 1892 from *Oreinus sinuatus,* fresh water fish in Kashmir. *Kashmir Sci.,* **5**(1-2):147-152.

Fotedar, D. N. (1968a). On two oxynrid nematodes of genus *Thelandros* Wedl, 1962 from some lizards of Jammu and Kashmir and notes on the genera, *Thelandros* and *Parapharyngodon* Chatterji, 1933. *Kashmir Sci.,* **5**(1-2): 208-221.

Fotedar, D. N. (1969). Note on some Georgoderid trematodes of fishes and amphibians in Kashmir and other parts of India, and notes on the classification of the family Gorgoderidae Loss, 1901. *Kashmir Sci.,* **6**(1-2):89-100

Fotedar, D.N.(1970).On a new trematode, *Pleurorchis nensie* gen et sp. Nov. from *Bufo viridis* in Kashmir. Proc. 57th Sess. Indian Sci. Congr., Part-III, pp. 454-455 (Abt.)

Fotedar, D.N. (1971). A review of the genera in Pleurogenetinae Looss, 1899 (Trematoda:Lecithodriidae). *Kashmir Sci.,* **8**(1-2):127-142.

Fotedar, D. N. (1980). *Oswaldcruzia kashmirensis* n. sp. (Nematode: Trichostrongylidae) from *Bufo viridis* in Kashmir. *Indian J. Zool.,* **8**(2):27-31.

Fotedar, D. N. & Bambroo, N. (1965). On a new Trichostrongylid Nematode from sheep in Kashmir. *Kashmir Sci.,* **2**(1-2):136-141.

Fotedar, D. N. & Bambroo, N. 1967 (1968). On a new species of the nematode genus *Chabertia* Railliet & Henry: 1909 from sheep in Kashmir. *Indian J. Helminthol;* **19**(2): 192-196.

Fotedar, D. N. & Bambroo, N. 1975 (1978). On a new species of cestode genus *Thysaniezia* Skrjabin, 1926 from local sheep in Kashmir. *J. Sci. Univ. Kmr.,* **3**(1-2):205-108).

Fotedar, D. N. & Chishti, M. Z. (1973). On a new species of *Anomotaenia kashmirensis* (Choanotaenidae Mathevossian, 1953) from *Sturnus vulgaris* in Kashmir. *J. Sci. Univ. Kmr.,* **1**(1-2):48-50.

Fotedar, D. N. & Chishti, M. Z. (1974). On *Toxocara mystax* (Zeder, 1800) Stiles, 1907 from *Felis domestica* in Kashmir. *J. Sci. Univ. Kashmir,* **2**(1-2):16-18.

Fotedar, D. N. & Chishti, M. Z. (1974a). Redescription of *Choanotaenia crioti* Joyeux et Baer, 1955 and *C. infundibutum* (Bloch, 1779), with a note on the synonymy of *C. dutti* Mukherjii. 1964, *J. Sci. Univ. Kmr.,* **2**(1-2):73-78.

Fotedar, D. N. & Chishti, M. Z. (1976). *Anomotaenia acrocephati* n. sp. and first record of *A. galbulae* (Gmelin, 1790) Fuhr., 1932 from some birds of Kashmir. *Rivista di Parasitologica,* **37**(2):247-252.

Fotedar, D.N. & Chishti, M.Z. (1980). *Psendoschistotaenia indica*, new genus, new species (Amabilcidae, Cestoda) from *Podiceps ruficollis capensis* in Kashmir, India. *Riv. Parassitol.*, **41**(1):39-44.

Fotedar, D. N. & Chishti, M. Z. (1980a). *Streptocara indica* n. sp. (Acuarioidae Sobolev, 1949: Nematoda) from *Gallus domestica* in Kashmir, India. *Indian J. Helminthol.,* **32**(1):4-7.

Fotedar, D.N., Chisthti, M.Z. & Khan, A.R. (1970). A new record of *Prosthogonimus (Macrogenotrema) cuneatus* (Rudolphi, 1809) from cloaca of *Passer domesticus* from Kashmir. *Kashmir Sci.,* **7**(1-2): 92-94.

Fotedar, D. N. & Dhar R. L. (1970). On the morphology of *Filochona kashmirensis,* (Thapar, 1950) from fresh water fishes of Kashmir. *Kashmir Sci.,* **7**(1-2): 107-110.

Fotedar, D. N. & Dhar R. L. 1973, (1975). On a species of nematode Genus *Spinitectus* Fourment from *Oreinus richardsonii,* a fresh water fish of Jammu, India. *J. Sci. Kash. Univ.,* **1**:63-66.

Fotedar, D.N. & Dhar, R.L. 1973(1975a). On the first record of a later juvenile form of an Acanthoccphalan Genus *Polymorphus* Luhe, 1911 from Rectum *of Rana Cyanophlyctis* in Kashmir. J.Sci. Univ. *Kashmir.,* **1**(1-2):55-56.

Fotedar, D.N. & Dhar, R.L. (1977). On five new species of genus *Pomphorhynchus* Monticelli 1905 from fresh water fishes of Jammu & Kashmir, India. *Indian J. Zool.,* **5**(1):15-24.

Fotedar, D.N., Duda, P.L. & Raina, M.K. (1970). A new species of the genus *Pomphorhynchus* Monticelli., 1905 from *Bufo viridis* in Kashmir. *Indian J. Helminthol.,* **22**(1):41-45.

Fotedar, D. N. & Handoo, Z.A. (1974). Two new species of *Helicotylenchus* Steiner (Hopololaiminae Nematoda) from Kashmir. *J. Sci. Univ. Kashmir*, 2(1-2): 55-62.

Fotedar, D. N. & Handoo, Z. A. (1977). *Aerotylenchus safroni* n. gen., n. sp. (Nematoda: Tylenchida) from Kashmir. *Indian J. Nematol.,* **7**(2):145-147.

Fotedar, D. N. & Handoo, Z., 1975 (1978). A revised scheme of classification to order Tylenchida Thorne, 1949. (Nematoda) *J. Sci. Univ. Kashmir,* **3**(1-2):55-82.

Fotedar, D. N. & Kaul, V. (1982). Plant-parasitic nematodes associated with various crops in Kashmir. *Kashmir Univ. Res. J.,* **2**:26-31.

Fotedar, D. N. & Kaul, V. 1983 (1984). *Tylenchus andrassyi* sp. nov. (Nematoda: Tylenchinae) from Kashmir. *J. Helminthol.* **35**(1):43-46.

Fotedar, D. N. & Kaul, V. (1985). On some species of *Helicotylenchus* Steiner, 1945. (Hoplomaimidae: Nematoda) as common plant parasitic nematodes in Kashmir, India. *Indian J. Nematol.,* **15**(1):9-13.

Fotedar, D. N. & Kaw, Lalita (1965). New Diplotriaenid nematode from the body cavity of *Acridotheres tristis* (Linnaeus). *Kashmir Sci.,* **2**:125.

Fotedar, D.N. & Khan, A.R. (1974). *Paralecithodendrium Indicum* n.sp. from the intestine of *Pipistrellus mimus*, common Bat in Kashmir. *J. Sci. Univ. Kashmir,* **2**(1-2):63-65.

Fotedar, D.N. & Khan, A.R. (1974a). *Plagiorchis (Plagiorchis) pipistrellicola* n. sp. from Kashmir Bat, *Pipistrellus mimus. J. Sci. Univ. Kashmir.,* **2**(1-2):66-69.

Fotedar, D.N. & Khateeb, N.G., 1986(1987). Occurrence and seasonal variations of Helminth parasites of domestic fowl in Kashmir. *Indian. J. Helminthol.,* **38**(1):49-54.

Fotedar, D. N. & Mahajan, R. (1971). On a new species of genus *Tylenchorhynchus* Cobb. 1913 (Nematoda: Tylenchorhynchidae) from Kashmir, India. *Kashmir Sci.,* **8**(1):120-122.

Fotedar, D. N. & Mahajan, R. (1972). A new species of the genus *Basiria* Siddiqi, 1954 from around the roots of *Brassica oleracea* in Kashmir *Proc. Indian Sci. Congr.,* Part III, p. 430.

Fotedar, D. N. & Mahajan, R. (1972 a). Two new nematode species (Nothotylenchidae) from Kashmir. *Indian J. nematod.,* **2** (2):169-172.

Fotedar, D. N. & Mahajan, R. (1973). On two species of the genus *Basiria* Siddiqi, 1959 (Nematoda: Psilenchinae) from Kashmir, India. *Indian J. Zool.*, **1**(1):49-52.

Fotedar, D. N. & Mahajan, R. (1974). Four new species of genus *Helicotylenchus* Steiner, 1945 (Nematoda: Hoplolaiminae) from India. *Rivista di Parasitologica,* **35**(2):119-124.

Fotedar, D.N., Mahajan, R., Dhar, R.L. & Chishti, M.Z. (1970). New variety of *Raillietina* from common Blue-Rock Pigeon in Kashmir. *Kashmir Sci.,* **7**: (1-2):103-106.

Fotedar, D. N. & Parveen, Q. N. (1987). On two species of trematode genus *Diplozoon* Nordmann, 1832 from fishes of Kashmir and notes on some species of the genus. *Indian J. Helminthol.,* **39**(1):56-76.

Fotedar, D.N. & Raina, M.K. (1964). On a new species of the trematode genus *Posthodiplostomum* Diabois, 1936 from *Milvus migrans lineatus* (Gray), common kite in Kashmir. *Kashmir Sci.,* **1**(1&2):64-69.

Fotedar, D.N. & Raina, M.K. (1965). On a new species of the trematode Genus *Lyperosomum* Looss, 1899 from *Corvus monedula soemmeringii* common Jackdew in Kashmir. *Indian. J. Helminthol.,***17**(1):54-62.

Fotedar, D. N., Raina, M. K., Mahajan, R. & Dhar. R. L. (1972). Redescription of *Contracaecum milvi* Yamaguti; 1940 from *Milvus, migrans lineatus* from Kashmir. *Indian J. Helminthol.,* **24**(1-2):1-4.

Fotedar, D. N., Raina, M. K. & Tikoo, Ratni (1966). On a new species of the nematode, Genus *Kalicephalus* Moliu, 1861, from a snake in Kashmir. *Kashmir Sci.,* **3**(1-2) : 92-100.

Fotedar, D. N. & Wali, N. 1982 (1984). On the development of juvenile stages of some bursate nematode parasites of sheep and pathogenecity of *Oesophagostomum. Indian J. Helminthol.,* **34**(2):123-135.

Gordh, G. & Headrick, D. H. (2001). *A Dictionary of Entomolgy.* CABI Publishing, CAB International, Wallingford, Oxon, U.K., 1032 p.

Gupta, N. K. & Uma (1980). *Quinisulcinus punici* (Tylenchorhynchidae: Tylenchida), a new nematode species from India. *Proc. Indian Acad. Sci.,* (Animal Sci.), **89**(5):415-418.

Gupta, N. K. & Uma (1981). Descriptions of two new species of genus *Tylenchorhynchus* Cobb, 1913, family Tylenchorhynchidae (Eliava, 1964). Golden, 1971 from India. *Helminthologica,* **18**(1):53-59.

Gupta, Rahul, Shah, Imtiyaz & Gupta, S. L. (2002). Echinococceal Disease of Kidney, presenting as renal failing defect. *JK* Practioner, **9**(4): 254-55.

Gupta, S.P. (1967). Helminthic-fauna of Kashmir. *Kashmir Sci.,* **4**(1-2):56-61.

Gupta, Santosh (1978). Discovery of genus *Echthrus* in India (Hymenoptera: Ichneumonidae). *Oriental Inst.,* **12**(2):211-214.

Gupta, V. K. (1967). Indian species of *Itoplectis* Forester (Hymenoptera: Ichneumonidae). *Oriental Inst.,* **1**(1-2):45-53.

Gupta, V. K. (1980). A revision of the Tribe Poemenini in the Oriental region (Hymenoptera: Ichneumonidae). *Oriental Inst.,* **14**(1):73-130.

Gupta, V. K. (1984). The Tribe Tryphonini in India, with descriptions of new species (Hymenoptera: Ichneumonidae). *Oriental Inst.,* **18**:173-186.

Gupta, V. K. (1985). A review of the tribe Neoxoridini of the world (Hymenoptera: Ichneumonidae Pimplinae). *Oriental Inst.,* **19**:323-329.

Gupta, V. K. 1984 (1985a). The Tribe Tryphonini in India with description of new species (Hymenoptera: Ichneumonidae). *Oriental Inst.,* **18**:173-186.

Gupta, V. K. (1991). A review of the exenterine Genus *Eridolius* (Hymenoptera: Ichneumonidae) and description of new species from Oriental region. *Oriental Insects,* **25**:435-446.

Gupta, V. K. & Chandra, Girish. (1974). Oriental species of *Xorides (Xorides)* (Hymenoptera: Ichneumonidae). *Oriental Inst.,* **8**(4):395-411.

Gupta, V. K. & Tikar, D.T. (1967). Indian species of *Scambus* Hertig. (Hymenoptera: Ichneumonidae). *Oriental Inst.,* **1**(3-4):215-237.

*Handoo, Z. A. (1977). *Soil and Plant Parasitic nematodes of vegetables and fruit crops in Kashmir.* Ph.D. Thesis, Deptt. of Zool., Faculty of Science, University of Kashmir, Srinagar, 290 pp.

Hangloo, V. K., Koul, I. Dev, G., Dhar, U., Chrungoo, R. K. & Khan, M. (1989). *Biliary* ascariasis. *Indian J. Gastroenterol.,* **8**(4):265-266.

Hansson, C. (1985). The entedontine genera – Achrysocharoides Girault, *Chrysocharis,* Foerster and *Kratoysma* Boucek (Hymenoptera: Eulophidae) in Oriental region. *Entomologica Scand.,* **16**(3):217-226.

Harwood, R.F. & Maurice, J.T. (1979). *Entomology in Human and Animal Health* (7th ed.). MacMillan Publishing Co. In., New York, U.S.A. 548 pp.

Hayat, M. (1972). The species of *Aphelinus* Dalman, 1820. (Hymenoptera: Aphelinidae) from India. *Entomophaga,* **17**(1):49-58.

Hayat, M. (1975). Some Indian species of *Anastatus* (Hymenoptera: Chalcidoidea: Eupelmidae). *Oriental Insects,* **9**(3):261-271.

Hoogstraal, H. & McCarthy, V.C. (1965). Host and distribution of *Haemaphysalis kashmirensis,* with description of immature stages and definition of the subgenus *Herpetobia canestrini* (resurrected). *J. Parasitol.,***51**: 674-79.

Hoogstraal, H. & Verma, M.G. R. (1962). *Haemaphysalis cornupunctata* and *H. kashmirensis* sp. n. from Kashmir, with notes on *H. sundrai* Sharif and *H. swelli* Sharif of Indian and Pakistan (Ixodoidea: Ixodidae). *J. Parasitol.,* **48:**185-194.

Hopkins, G. H. E. & Rothschild, M. (1953, 1962, 1966, 1966a, 1971). *An illustrated Catalogue of the Rothschild Collection of fleas (Sphonaptera) in the British Museum (Natural History).* Vol. I., (1953), 361 p. *Tungidae and Pulicidae.*, Vol. II, (1962), 445 p., *Vermipsyllidae to Xiphinopsyllidae;* Vol., III, 500 p. (1966), *Histrichopsyllidae;* Vol. IV, p. 549 (1966a), *Histrichopsyllidae* and Vol.V, 530 pp. (1971), Leptopsyllidae. British Museum, National History, London, England.

Horstmann, K. (1992). Neue Taxa dev campopteginae aus deu Gattungen *Campoplex* Gravenhost *Daidegma* Forster and *Nemeritis* Holmgren (Hymenoptera Ichneumonidae). *Z. Arbeitsgem Oesterr. Entomol.,* **44**(3-4):116-127.

Husains, Tasawer & Agarwal, M. M. (1981). Taxonomic studies on Indian Epitraninae (Hymenoptera: Chalcididae). *Oriental Insects.,* **15**(4):413-432.

Hyman, L.B. (1951). *The Invertebrates,* Vol. III, McGraw Hill, New York, U.S.A.

Iyenger, R. (1973). The Siphonaptera of Indian sub-region *Oriental Insects* (Suppl.), **3**(1-102).

Jacob, V. (1950). *Indian J. Malariology,* **4**: 251-260 (*cf.*)

Jacob, V. P. & Kalra, S. L. (1951). Kalazar in Kashmir. *Indian J. Med. Res.,* **39**: 323-327.

Jairajpuri, M. S. & Baqri, G. H. (1966). *Sectonema procta* n. sp. and *Pungenatus angulatus* n. sp., two new soil inhabiting nematodes. *Nematologica,* **12**(3):396-402.

Jairajpuri, M. S. & Loof, P. L. A. (1976). On some species of *Euchodelus* Thorne, 1933. (Nematoda: Dorylaimidae) from India. *Nematol.,* **12**(3):501-508.

Jairajpuri, M. S. & Siddiqi, A. H. (1963). A new and known species of the genus *Tylencholaimellus,* M.O. Cobb., 1915 (Nematoda: Dorylaimoidea) from India, with key to its species. *Zeitschrift fur Parasitenkunde,* **22**(6):489-494.

Jairajpuri, M. S. (1965). *Basiria kashmirensis* n. sp. (Nematoda: Tylenchida) from India. *Labdev J. Sci. Technol.,* **3**(1):23-25.

Jordan, E.L. (1930). On a new species of *Lepoderma* (Looss) *Lepoderma himalayii* n. sp. *Proc. Ind. Sci. Cong.,* **17**:246.

Kalra, S. L. (1952). Natural History of Typhus fever in India. *Indian J. Med. Sci.,* **6**: 569-575.

Kalra, S. L. (1959). Progress in the Knowledge of Rickettsiae Disease in India. *Indian J. Med. Res.,* **476**: 477-483.

Kalra, S. L. & Rao, R. N. A. (1949). Typhus fever in Kashmir State. Part I. Epidemic Typhus. *Indian J. Med. Res.,* **37**(4): 395-400.

Kalra, S.L. & Rao, R.N.A. (1951). Observation on the epidemiology of relapsing fever in Kashmir. *Indian J. Med. Res.,***39**:319-21.

Kalra, S.L. & Rao, R.N. A. (1951a). Typhus fever in Kashmir State Part II. Murine Typhus *Indian. J. Medical Res.,* **39**(3):297-302.

Kani, A., Nagase, K., Totani, Morishita, T. & Okamoto, K. (1981). An observation of parasitic and sanitary animals in Jammu & Kashmir State of India. 51 (Japanese) (cf. *Helminthol.* Abt. Vol. 50(6): 325 p.).

•Kaul, Angali (1987). *Studies on Protozoan parasites of some Invertebrates in Kashmir.* M.Phil. Dissertation. P.G. Deptt. of Zool. Faculty of Science, University of Kashmir, Srg.

Kaul, H. N., Kulkarni, S. M., Dhanda, V. & Geevarghese, G. (1978). *Huntrellus sagarensis,* parasitizing *Haemaphysalis bispinosa* (Acari: Ixodida) in Jammu & Kashmir, India. *Acrology Newsl.,* **6**:4-5. (hymenopteran parasite, newly recorded as parasite of tick).

Kaul, H. N., Shietty, P.S., Ghalasi, G.R. & Dhanda, V. (1990). Survey of ticks (Acarina: Ixodidae) for Crimean haemorrhagic fever virus activity in Jammu & Kashmir State, India. *Indian J. Medical Res.,* Section A, *Infections Diseases,* **91**(Jan.):5-8.

*Kaul, Ramesh (1985). *Studies on the nematode parasites of Brassicaceous plants of Kashmir*, Ph. D. Thesis P. G. Deptt. of Zool., Faculty of Sci., University of Kashmir, Sgr., Kmr. (J&K). 338 p.

Kaul, S. (1949). Relapsing fever tick-borne on account of an outbreak in Jammu & Kashmir forces, India. *Indian Med. Gaz.,* **84**:433-440.

Kaul, S. N. (1944). *Report of typhus outbreaks at Banihal submitted to D. M. S.,* J&K (Govt.).

Kaul, V. (1984). Seasonal fluctuation of phytonematodes in two fruit nurseries in Kashmir. *Indian J. Helminthol.,* **37**(1):4-11.

Kaul, V. (1984a). *Studies on the Plant-parasitic Nematodes in Kashmir*. Ph.D. Thesis, P. G. Deptt. of Zool., Faculty of Science, Univ. of Kmr, Srinagar, Kmr. (J&K), 247 pp.

Kaul, V. (1987). Two new species in the super-family Hopolaimoidea (Nematoda) from Kashmir. *Geobios New Reports,* **6**:162-164.

Kaul, V. (1998). Plant-Nematology in Jammu and Kashmir: Aspects and Prospects, pp.18-22. *In:* Trivedi, P. C. (Ed.), *Phytonematology in India.* CBS Publishers and Distributors, Daryaganj, New Delhi, 205 p.

Kaul, Vijay; Masoodi, M. A., Bhagat, K. C. & Bhat, O. K. (2000). Biology, behaviour and bio-control agents of *Drosicha dalbergia* (Grien), infesting Almonds in Kashmir. *Ann. Plant. Prot. Sci.,* **8**(1):24-26.

Kaul, V & Fotedar, D.N. (1987). Morphometric variability in *Helicotylenchus erythrinae* (Eimmermann)1904 Golden 1956. *Geobios New Rept*., 6:138-142

Kaul, V. & Wali-ullah, M. I. S. (1989). A new species of *Pratylenchus* Micotetzky, 1922 (Pratylenchidae: Nematoda) from Kashmir. *Indian J. Nematol.,* **19**(1):66-69.

Kaul, V. & Wali-ullah, M. I. S. (1995). *Tylenchorhynchus oryzae* sp. n. (Tylenchorhynchidae: Nematoda) from Kashmir, India. *Annals. Plant Protection Sci.,* **3**(2):155-157.

Kaw, B.L. (1941). Studies on the Helminth parasites of Kashmir. Part I. Descriptions of some new species of genus *Pomphorhynchus* Monticelli (1905). *Proc. Indian Acad. Sci.,* **13**:369-378.

Kaw, B.L. (1943). Studies on the helminth parasites in Kashmir. Part II. On two new trematodes of the sub-family Pleurogenitinae looss, 1899, with a review of the genus *Pleurogenes* Looss, 1896. *Proc. Indian Acad. Sci.,* **18**(3), Sec. B: 97-108.

Kaw, B. L. (1944). Studies on helminth parasites of Kashmir. Part III. Description of a new Allocreadid, *Crepidostomum indicum* from a fresh water fish, *Schizothorax niger* from Dal lake, Kashmir. *Proc. Ind. Acad. Sci.,* **20**:72-77.

Kaw, B.L. (1945). On the present status of the genus. *Loxogenes. Proc. Indian Acad. Sci.,* **21**(6):342-43.

Kaw, B.L. (1950). Studies in Helminthology:Helminth parasites of Kashmir. Part-I. Trematodes. *Indian J. Helminthol.,* **2**(2) :67-126.

Kaw, B.L. (1951). Studies on Helminthology. Helminth Parasites of Kashmir. Part II. Acanthocephala. *Indian. J. Helminthol.,* **3**(2):117-132.

*Kaw, Lalita (1968): *Studies on Helminth Parasites of some vertebrates in Kashmir: Helminth parasites of some birds in Kashmir,* Ph. D. Thesis. P. G. Deptt. of Zool., Faculty of Sci., Univ. of Kmr., Sgr., Kmr. (J&K).

Kettle, D.S. (1995). *Medical and Veterinary Entomology* (2nd edition): CAB International, Willingford Oxon, U.K., 725 p.

Khajuria, R.R., (1966). Incidence and control of parafilariasis in bovine in Jammu Province. *Indian Vet. J.,* **43**(8):749-50.

*Khan, A. R. (1974). *Studies on the trematode parasites of some vertebrates of Kashmir.* Ph. D. Thesis P. G. Deptt. of Zool., Faculty of Sci., Univ. of Kmr., Sgr., Kmr. (J&K).

Khan, A.R. 1986(1987). On a new trematode, *Opisthorchis anasi* sp. nov. from Kashmir. *Indian J. Helminthol.,* **38** (2):119-122.

Khan, A. R. & Ahmad, D. 1991 (1992). Redescription of *Hypoderaeum conoideum* (Bloch, 1982), Dietz, 1909 from *Anas platyorhynchos domesticus* in Kashmir. *Indian J. Helm.,* **43**(2):162-164.

Khan, A.R & Ahmad, F. (1992a). A new record of *Echinoparyphium chinense* Lu. Li et Chu,1964 from *Anas platyrhynichos domesticus* in Kashmir. *Indian J. Helm.,* **44** (2):146-148.

Khan, A.R. & Chishti, M.Z. (1982). On *Echinostoma revolutum* (Froelich) Loosse, 1899 and synonymy of *Neoechinostoma spinosa* Agarwal, 1963. *Kashmir Univ. Res. J.,* **2**:22-24.

Khan, A.R. & Chishti, M.Z. (1983). Studies on *Plagiorchis (Plagiorchis) tehri* n.sp. from *Passer domesticus* from Bachpora, Kashmir, *J. Indian, Instit. Sci.,* **64**(12):189-192.

Khan, A.R. & Chishti, M.Z. (1984). *Euparyphium pindchi* sp. nov. from a common Grebes, *Podiceps ruficollis capensis. Indian J. Helminthol.,* **36**(2):134-138.

Khan, A.R. & Chishti, M.Z. 1987(1989). The incidence of trematode parasites in a common Kashmiri fowl population and study of *Catatropis kashmirensis* n.sp. (Digenea:Notocotylidae Luhe, 1909). *Riv. Parasitol.,* **48** (3):355-360.

Khan, A. R., Perhar, A. K., Sharma, D. B. & Lahari, V. C. (1975). Kala-azar in Jammu. *J. Indian Med. Res.,* **65**(3): 85-86.

Khan, A.R. & Rasheed, Majidah 1993 (1994). Influence of season on the intensity of helminth infection in *Schizothorax esocinus* Heckel from the Wular Lake. *Indian J. Helminthol.,* **45**(42):64-73.

Khan, A.R. Rasheed, Majidah & Kaul, J.A. (2002). Prevalence of Giardiasis in Kindergarten School Children of Urban Kashmir. *J. Parasitic Diseases,* **26** (1):30-33.

Khan, A. R., Rasheed, Majidah & Mir, Tajamul, H. (2003). A study on Mallophagon lice and some helminthes of Whooper Swan, *Cygnus Cygnus* of Kashmir. *Proc. Smr. on Interdisciplinary Approaches* of *Zoological Research*, held at Deptt. of Zool. Univ. Kmr., Sgr. Kmr. (J&K), from 29th September – 1st October, 2003, p.15 (Abt.).

Khan, E. & Siddiqi, M.R. (1963). *Criconema laterale* n. sp. (Nematoda: Criconematidae) from Srinagar, Kashmir. *Nematologica,* **9**:584-86.

Khan, M. D. (1985). *Studies on Amphistomes of Domestic Ruminants in Kashmir*. Ph. D. Thesis. P. G. Deptt. of Zool., Faculty of Sci., Univ. of Kmr., Sgr., Kmr. (J & K)

Khuroo, M.S. (1996). Ascariasis. *Gastroenterol. Clin.North America*, 25:553-57

Khuroo, M.S., Zargar, S.A. Yattoo, G.N. Dar, M.J. Javid, G., Khan, B.A. Boda, M.I. & Mahajan, R. (1992). Sonographic findings in gall bladder ascariasis *J. Clinical Ultra Sound,* **20**(9):287-91.

*Khar, H. K. (1983). *Studies on the larval ecology of trematods in some waterbodies of Jammu*. Ph. D Thesis. Department of Zoology, Jammu University, Jammu (J & K).

Kharoo, V.K. & Dhar, R.L. 1981(1982). On a new species of Trematoda, *Leucochloridium gallinuli* n.sp. from the Indian Moorhen, *gellinula chloropus* from Kashmir. *Indian. J. Helminthol.,* **33**(1):15-19.

Khuroo, M. S. (1999). Hepato-biliary & Pancreatic ascariasis in endemic areas. *JK – Practioner,* **6**(2):189.

Khuroo, M.S., Dar, M.Y., Yattoo, G.N., Zargar, S.A., Javaid, G.K., Khan, B.A. & Boda, M.I. 1993. Precutaneous drainage versus albendazole theraphy in hepatic hydatidosis:A prospective randomized study. *Gasteroenterology*, **104**:1452-1459.

Khuroo, M.S., Datta, D.V. Khoshy, A., Mitra, S. K. & Chuttani, P. N. (1980). Alveolar hydatid disease of the liver, with Bud-Chiari syndrome. *Post-graduate Medical J.,* **653**: 197-201.

Khuroo, M.S. & Zargar, S.A. (1985). Biliary ascariasis: Common cause of biliary and pancreatic disease in an endemic area. *Gastroenterology,* **88**(2):417-423.

Khuroo, M. S., Zargar, S.A., Koul, P., Khan, B. A., Dar, M. Y. & Alai, M. S. (1992 *a*). Ascaris-induced acute pancreatitis. *British J. Surgery,* **79**(12):1335-1338.

Khuroo, M. S., Zargar, S.A. & Mahajan, R. (1991). *Echinococcus granulosus* cysts in liver: Management with preceutaneous drainage. *Radiology,* **180**(1): 141-145.

Khurma, Uma, R. (1988). Plant-parasitic nematodes from some hilly areas of North-India. *Indian J. Nematol.,* **18**(1): 132-133.

Koul, P.A., Waheed, A., Hayat, M. & Sofi, B.A. (2000). Praziquantel in nidosamide resistant *Taenia saginata,* infection. *Scandinavian J. Infectious Diseases*, **31**(6):603-604.

Koul, P.L., Raina, M.K. Bambroo, P. and Koul, U. 1991(1992) *Pallisentis jagani* sp. nov. from *Channa channa* in Jammu, India. *Indian J. Helminthol.,* **43**(2):124-28.

*Koul, Ravinder, Kumar (1977). *Studies on helminth parasites of Reptiles of Jammu & Kashmir.* Ph. D. Thesis. P. G. Deptt. of Zool., Faculty of Sci., Univ. of Kmr., Sgr., Kmr. (J&K). 315 pp. & LXXX plates.

Koul, Tej, Krishni & Raina, M.K. (1989). A new copepod parasite, *Lernaea gambusiae* n.sp. (Lernaeidae. Lernaeninae) infecting the fish, *Gambusia affinis* in Kashmir. *Mysore J. Agric, Sci.,* **23**(4):543-548.

Krishna-Prasad, K. S. (1986). Plant-parasitic nematodes of India, Problem and progress *In:* Swarup, G. & Das – Gupta, D. R. (Eds.), Indian Agricultural Research Institute, New Delhi.

Kulkarni, S. M. & Bhat, H. R. (1972). *Neopsylla kashmirensis,* sp. nov. from Kashmir, India. (Siphonaptera: Hystrichopsyllidae). *Oriental Insects,* **6**(4): 513-519.

Kullandar, S.O., Fang Fang Delling, B. & Ahlander, E. (1999). *Fishes of the Kashmir Valley*. pp.99-167, *In:* Nyman, L. (Ed.). *River Jehlum, Kashmir Valley Impact on the Aquatic Environment*. Published by Swedmar. The International Consultancy Group of the National Board of Fisheries, Sweden.

Kumar, K., Rahman, S. J., Nim, S., Chirwatkar, K. R. & Kalra, N. L. 1983 (1984). A note on sandflies of Jammu. *J. of Communicable Dis.,* **15**(4): 287-289.

Lewis, D. J. (1978). The Phlebotomine sandflies (Diptera: Psychodidae) of the Oriental region. *Bull. British Mus.* (Nat. Hist.), **37**(6): 219-243.

Luc, M. & Goodey, J. B. (1963). *Hirshmanniella* nom nov. for *Hirschannia Nematologica,* **9**:471.

Luc, M., Sikora, R. A. & Bridge, J. (Eds.) (1990). *Plant-Parasitic Nematodes in sub-tropical and Tropical Agriculture*. CAB International Willingford, U. K., Oxon.

Mahajan, R. (1970). Effect of light on the surviual of *Hopolaimus* sp., the soil nematode surrounding the roots of *Zea mays* in Kashmir. *Indian J. Helminthol.,* **22**(1):36-40.

Mahajan, R. (1971). A preliminary report on the occurrence of plant-parasitic and soil nematodes in Jammu and Kashmir State. *Kashmir Sci.,* **8**(12):115-119.

Mahajan, R. (1972). On two soil nematodes, *Discolaimoides indicum* sp. nov. and *Pungentus juglensi* sp. na. from around roots of *Juglans regia* in Kashmir. *J. Helminthol.,* **46**(2):161-165.

Mahajan, R. (1972a). *Studies on the soil and plant-parasitic Nematodes in Kashmir.* Ph. D. Thesis, Deptt. of Zool., Faculty of Sci., Univ. of Kmr., Sgr., Kmr.

Mahajan, R. (1973). Two new nematodes, *Tylenchus kashmirensis* n. sp. and *Dorylaimoides pari* n. sp. from Kashmir. *Rivista Parasit.,* **34**(3):201-204.

Majidah, Rasheed Khan, A.R. (1998). Seasonal population dynamics of some helminth parasites of fishes in Wular Lake, Kashmir. pp. 34-40, *In: Parasitic Diseases-New Horizon* (*Proc.* 8th Natl. *Congr. Vet. Parasitol* and *Natl. Symp. On Molecular parasitol.,* held at CCS Haryana Agr. Univ., Haryana, from 9-11 Oct. 1998.

Makauer, M. (1972). The aphid attacking genera of Aphelinidae (Hymenoptera), including new genus. *Canadian Entomologist,* **104**(11):1771-79.

Makhdoomi, D. M., Nasreen, S., Bandy, S. D. & Moulvi, B. A. (1995). Incidence of different ovine gastro-intestinal parasites of Kashmir. *Indian Vet. J.,* **72**(8): 898-900.

Malhotra, Y. R. & Joyti, M.K. (1972). A new copepod parasite, *Lernaea kashmirensis* n. sp. (Lernaeidae, Lernaeinae) infecting stone loach in Kashmir. *Vest.* Cs. *Spol. Zool.,* **36**(2):119-122.

Malik, G.M., Romshoo, G.J., Hussain, Ashaq, Ahmad, Zubair, Basu, J.A. & Bhat, G.H. (1997). Hydatid Disease of liver, presenting as cholangitis. *JK-Practioner*, **4**(3):194-96.

Malla, A. K. (1981). *Morphology, Histology and Histochemistry of some cestodes.* Ph. D. Thesis. P. G. Deptt. of Zool., Faculty of Sci., Univ. of Kmr., Sgr., Kmr. (J&K).

Mani, M. S. & Kaul, B. K. (1973). On some chalcidoidae from India. *In:* Mani, M.S., Dubey, O.P. Kaul, B. K. & Saraswat, C. G. *Mem. School. Entomol.* St. John College, Agra, U.P.

Masoodi, M. A; Bhagat, K. C. & Sofi, M. R. (1996). Preliminary observations on the natural enemies of San Jose Scale, infesting apple trees in Kashmir. *Shashpa,* **3**:77-78.

Masoodi, M. A. & Trali, A. R. (1987). Seasonal history and biological control of San Jose Scale, *Quadraspidiotus perniciosus* Comstock (Diapididae: Homoptera) on apple in Kashmir. *J. Biological Control,* **1**:3-6.

Masoodi, M. A. & Trali, A. R. (1987a). New records of Scale Insects (Diaspidae: Homoptera) from Kashmir. *Entomon.,* **12**(3):227.

Masoodi, M. A., Trali, A. R. & Bhat, A. M. (1986). Further observations on the natural enemies of *Lymantria obfuscata* Walker (Lymantriidae: Lepidoptera) in Kashmir.*Entomon,* **11**(4):251-254.

Masoodi, M. A., Trali, A.R., Bhat, A.M., Tickoo, R.K. & Nehru, R.K. (1986a) Incidence of *Lymantaria obfuscata* (Lymantridae: Lepidoptera) in Kashmir.*Entomophaga*, **31**(4):401-404.

Masoodi, M. A., Trali, A. R., Bhat, A. M.., Tikoo, R. K. & Nehru, R. K. (1989). Establishment of *Aphytis* sp. *Proclia* group on San Jose Scale in Kashmir. *Indian J. Plant. Prot.,* **31**:71-73.

Masoodi, M. A., Trali, A. R., Bhat, A. M., Tikoo, R. K. & Nehru, R. K. (1989a). Establishment of *Encarsia* (*Prospaltella perniciosi*), a specific parasite of San Jose Scale, on apple in Kashmir. *Entomophaga,* **34**(1):39-43.

Mathur, A. C. & Sharma, M. L. (1962). *Hadronotus* sp. (Scelionidae, Proctrotrupoidae) as a parasite of eggs of *Cletus bipunctatus* West (Corecidae: Hemiptera). *Indian J. Ent.,* **23**(4):303.

Mathur, R. N. (1956). A new species of *Anastatus* Motschulsky (Chalcidoidea, Eupelmidae) from Kashmir. *Proc. R. ent. Soc. London* (B), **25**:93-97.

Mathur, V. K. & Prasad, S. K. (1971). Occurrence and description of *Hirschmanniella oryzae* in Indian Union, with description of *H. mangoriensis* n. sp. *Indian J. Nematol.,* **1**(1):220-225.

Mehra, H. R.(1937). Certain new and already known distomes of the family Lepodermatidae Odhuer (Trematoda) with discussion on the classification of the family. *Zeitschr. Parasit.,* **9**:429-469.

Mehta, U. R. & Raski, D. J. (1971). Revision of genus *Criconema* Hofmanner and Menzel, 1914 and other related genera (Criconematidae: Nematodes). *Indian J. Nematol.,* **2**:145-198.

Mir, A. S., Pandit, B. A., Shahardar, R. A. & Banday, M. A. A. (1992). Prevalence of ectoparasites in indigeneus fowls of Kashmir Valley. *Indian Vet. J.,* **70**(1):1071-1072.

Mir, A.S., Pandit, B.A., Shahardar, R.A. & Banday, M.A. (1993) Prevalence of ectoparasites of indigenous fowl of Kashmir. *Indian Vet. J.,* **70**(11):107-172.

Mir, Abdul Aziz (1992). *Studies on the Parasitoid complex of Galliform, Anseriform and Podicipediform Birds of Kashmir.* Ph. D. Thesis. Deptt. of Zool., Faculty of Sci., Univ. of Kmr., Srg, Kmr. (J & K).

Mir, I.A., Wani, N.A. Ahangar, A.G. Shah, O.J., Saleem, K. & Patnaik, Rekha (2002). Surgical ascariasis in children. *JK-Practioner,* **10**(1):17-21.

Mir, M.A., Raina, Puran, Matto, T.K. & Sethi, A.S. (1979). Incidence of Giardiasis in Kashmiri Childrens. *Indian. J. Pediatr.,* **46**:358-361.

Misger, M.S. & Mufti Mahmood (2000). Hydatidosis: A surgical jargon. *Surgery,* **5** (6):37-38.

Mishra, A. C. (1981). The hoplopleurid lice of Indian sub-continent. *Rec. Zool. Survey India.* (Mis. Pub.), Occasional Pap. No. 21. Issued by Director Zoological Surv. of India.

Mishra, A. C., Kulkarni, S. M. & Bhat, H. R. (1973). *Hoplopleura himalayana,* sp. n. (Anoplura: Hoplopleuridae), parasitizing *Apodema flavicollis* in the Himalayan region of India. *Oriental Insects.,* **7**(4): 501-506.

Mitra, R. D. (1959). Notes on sandflies of Punch and Riasi Districts of Kashmir. *Z. Tropenmed Parasit.* **10**(1): 56-66.

Mitra, R. D. & Roy, D. N. (1953). Notes on sandflies Part III. *T. Tropenmed Parasit.,* **88**: 324-326.

Mohan Rao, C.V.R. (1986). Prevalence of Crimean haemorrhagic fever Congo-virus in Jammu & Kashmir State. *Indian J. Med. Res.,* **84**:134.

Morley, Claude (1913). *The fauna of British India, including Ceylon. Hymenoptera – Vol.III. Ichneumonidae. Ichneumones* Deltoidei. Taylor & Francis, London, England.

Munshi, N. A., Dar, G. H., Shawl, A. S., Amman-ullah & Dar, G. H. (2001). Report on occurrence of some mushroom nematodes in Jammu and Kashmir. *Appl. Biol. Res.,* (CAAS, Shalimar, Kashmir), **3**:64.

Nair, C. P. (1973). Malaria in Kashmir Province of Jammu & Kashmir State – India. *J. Communicable Dis.,* **5**(1): 22-46.

Nandi, N.C. (1986). *Leucocytozoon squamatus* sp. n from scaly-bellied green wood-peeker, *Picus squamatus squamatus* Vigora. *Acta Protozool.,* **25**(2):223-226.

Nandi, N.C. & Mandal, A.K. (1978). Studies on some avian haematozoa from Jammu and Kashmir. *Bull. Zool. Surv. of India.,* **1**(2):103-106.Kaul, Anjali (1987). *Studies on Protozoan Parasites of some Invertebrates in Kashmir.* M.Phil. Dissertation Faculty of Sci., Univ. of Kashmir, Srinagar.

*Narain, Kanwar (1993). *Protozoan and Platyhelminth parasites of some aquatic vertebrates of J&K.* Ph. D. Thesis, Faculty of Sci., P. G. Deptt. of Zool., The Univ. Kmr., Sgr., Kmr. (J&K).

Narain, Kanwar & Raina, M.K. (1992). A new report of 'White spot disease' in fresh-water fish of Jammu and Kashmir, with a note on its histopathology. pp.115-120., *In: Proc. Curr. Trends in Fish and Fisheries Biology and Aquatic Ecology (Proc. 3rd. Natl, Semr on Fish and their environment,* held at Srinagar from 19-22 September, 1989. P.G. Deptt. of Zool. The Univ. of Kashmir, Srinagar, Kashmir.

Narayanan, E. S. (1957). Phenomenon of insect parasitism and their practical utilization in Biological Control of insect pests. *Proc. 44th. Sess. of Indian Sci. Congr.* **2**:23.

Narayanan, E. S. (1961). New records and description of new species of San Jose Scale parasites *Proc. Natl. Inst. Sci.*(B) *Silver Jublee* Suppl. 1960 (B): **26**:20-29.

Narayanan, E. S. & Lal, Kundan (1954). Studies on Indian Ichneumonidae (Hymenoptera) Part-II, Sub-family Pimplinae, Tribe Pimplini. *Indian J. Entomol.,* **16**:282-296.

Narayanan, E. S. & Lal, Kundan (1955). Studies on Indian. Ichneumonidae (Hymenoptera) Part-IV, Sub-family Pimplinae, Tribe Ephialtini, *Indian J. Entomol.,* **17**.

Narayanan, E. S., Rao, V.P. and Subba Rao, B.R. (1967). Advances made in Insect Parasitology and biological control of Pests in India. *In*: *Agricultural Entomology* ICAR, New Delhi. Pp.205-246.

Narendran, T. C., Buhroo, A. A. & Chishti, M.Z. (2001). Taxonomic studies on four new species of Chalcidoidea (Hymenoptera) of economic importance. *Entomol.,* **26**:147-159.

Narsimahan, D. (1962). Field survey of insecticide susceptibility of diseae vectors. II. *A. Pediculus humanus corporis* in Srinagar. *Armed Forces Med. J. India.* **18**: 418-419.

Niewiadomska, K. (2002). Superfamily Diplostomoidea Poiner, 1886. pp.159-196. *In:* Jones, A. & Bray, K.A.(Eds.) International, Wallingford, Oxon, U.K.

Pampori, R. A., Wani, Asif, Latoo, M. & Hakeem, A. (2003). Hydatid cyst in the neck – An unusual presentation. *JK-Practioner,* **10**(2): 130-131.

Pande, B.P. (1937) *Prostocus himalayai* n.sp., a frog trematoda (Lecithodendriidae). *Proc. Nat. Acad. Sci.,* **7**:202-204.

Pande, B. P. (1938). The trematode genus *Allocreadium* in North India fresh water fishes. *Proc. Ind. Acad. Sci.,* **7**:54-60.

Pandit, B.A., Mir, A.S. & Bandy, M.A.A. & Shahardar, R.A. (1991). Prevalence of Helminth parasites in indigenous fowls of Kashmir Valley *Poultry Advisor,* **24**:37-39.

*Pandoh, B.R.(1992). *Ecological studies on trematodes of some aquatic vertebrates of Jammu Province*, Ph.D. Thesis. Department of Zoology, Jammu University, Jammu (J&K)

Papp, J. (1985). Three new *Blacus* Nees, 1818 species from India (Hymenoptera: Braconidae: Blacinae). *Annales Hist. Nat. Mus. Natn. Hung.,* **77**:227-232.

*Parveen, Qazi Naz (1982). *Trematode parasites of fishes of Jammu & Kashmir State.* Ph. D. Thesis. P. G. Deptt. of Zool., Faculty of Sci., Univ. of Kmr., Sgr., Kmr. (J&K).

Patheja, J. S. (1963). Investigation Report of an epidemia of louse-borne typhus among Tibetan Rufugees at Leh (Eastern Ladakh) during July, 1961. *Armed Forces Med. J.* (India), **19**(3): 120-124.

Pawar, A. D. & Parry, M. (1989). Record of natural enemies of important fruit pests in Ladakh (J&K). *Ind. J. Plant Prot.,* **17**:291-293.

Pawar, A. D., Tuhan, N. C., Balasubramanium, S. & Parry, M. (1981). Biological control of codling moth in Ladakh. *Indian J. Plant Prot.,* **8**(2):189-191.

*Peerzada, M. Y. (1997). *Population dynamics of helminthes of some snow-fed streams of Kashmir.* Ph. D. Thesis P. G. Deptt. of Zool., Faculty of Sci., Univ. of Kmr., Sgr., Kmr. (J&K).

Petrotschenko, V.I. (1956). *Acanthocephala of domestic and wild animals.* J.I., 430 p.

Pojmanska, T. (2002). Family Brachylaimidae Joyeux and Foley, 1930. pp.32-39. *In.* Gibson, David, Jones A.J. and Bray, R.A. (Eds.), Keys to the trematodes (Vol.I). CBI Publishing CAB (International Willingford, Oxon, U.K.

Punjabi, A.A. & Saraf, S.K. (1969). Occurrence of *Varroa jacobsonii* in Kashmir. *Indian Bee J.,***3144**.

Puri, I. M. (1932). Studies on Indian Simuliidae. Pat 2. Descriptions of males, females and pupae of *Simulium rufibasis* Brunetti, its variety *fasciatum* nov. var. and of three new species from the Himalayas. *Indian J. Med. Res.,* **19**:899-915.

Puri, I. M. (1932a). Studies on Indian Simuliidae. Part. B. Description of males, females and pupae of *S. griseifrons* Brunetti (1911) and of four new species, with striped thorax. *Indian J. Med. Res.,* **19**:1125-1143.

Puri, I. M. (1948). The distribution of Anopheline mosquitoes in India, Pakistan, Ceylon and Burma. Part V, Additional Records, 1963-48. *Indian J. Malar,* **2**:67-107.

Putatunda, B.N. & Abrol, D.P. (2003) Mites associated with bees in Jammu & Kashmir. *Zoo's Print. J.,* **18**(2) (Coimbtore, India).

*Raina, M. K. (1968). *Studies on Helminth Parasites of some vertebrates in Kashmir. Monographic study* on *Avitellina* Gough, 1911 & Studies on *nematode parasites of shrew and Rat in Kashmir.* Ph. D. Thesis. P. G. Deptt. Of Zool., Faculty of Sci., Univ. of Kmr., Sgr., Kmr. (J&K).

Raina, M. K. (1970). On morphology of *Ganguleterakis spumosa* (Schneider, 1866) Lane, 1914 from the intestine of common rat, *Rattus rattus)* in Kashmir. *Kashmir Sci.,* **7**:95-98

Raina, M.K.(1971). On the morphology of *Prosthogonimus putschkowskii* Skrjabin,1912 from egg of domestic hen in Kashmir. *Zool. Anz.,***187**(5&6):324-327

Raina, M. K. (1973). Incidence of infection of sheep and goat with *Avitellina* Gough, 1911, *Stilesia* Railliet 1983 and *Moniezia* Blanchard 1891, in Kashmir. *J. Sci. Univ. Kmr.* **1**(1-2):59-62.

Raina, M. K. (1975). A monograph of the genus *Avitellina* Gough, 1911. (Avitellinidae: Cestoda). *Zool. Jb. Syst.,* **102**:508-552.

Raina, M. K., Chishti, M. & Koul, P. L., 1975 (1977). A new report of *Oochoristica tuberculata* (Rudolphi, 1819) Luhe, 1898, from the intestine of *Agama tuberculata* in Kashmir, with remarks on its synonyms. *Indian J. Helminthol.,* **29**(1):1-4.

Raina, M. K., Chishti, M. Z. & Khan, A. R. (1973). An abnormality in *Oochoristica tuberculata* (Rudolphi, 1812) Luhe, 1898. *J. Sci. Univ. Kashmir,* **1**:57.

Raina, M. K. & Dhar, R. L. (1972). On *Camallanus fotedari* n. sp. (Nematoda: Spiruidea) from the intestine of *Nemachilus kashmirensis* in Kashmir, Indian., *J. Helminthol.,* **46**(2): 157-160.

Raina, M. K. & Kaul, R. K. (1974). A new report of *Trichocephalus muris* Schrank, 1788 from *Rattus* rattus in Kashmir. *J. Sci. Univ. Kashmir,* **2**(1-2):70-72.

Raina, M. K. & Kaul, R. K. (1982). On a new species of *Capillaria* (Nematoda) from a bat in Kashmir. *Angew Parasitol.,* **23**:28-31.

Raina, M.K. & Khan, M.D. (1982). Prevalence of Amphistomes in Ruminants in Kashmir. *Kashmir Univ. Res. J.,* **2**:17-21.

Raina, M.K. & Khan, M.D. (1983). Excysted Amphistome metacercariae. *Indian J. Parasitol.,* **7**(1):115-116.

Raina, M.K. and Koul, P.L. (1983). The histopathology of *Neoechinorhynchus hutchinsoni* infection in *Nemacheilus kashmirensis. J. Helminthol.,* **58**:165-168.

Raina, M. K., Malla, A. K. & Raina, Usha 1975 (1978). On *Mathevotaenia sysmmetrica* (Baylis, 1927) Akumyan, 1946, from the intestine of *Ratus rattus* in Kashmir. *J. Sci. Univ. Kmr.,* **3**(1-2): 113-117.

Raina, M.K. & Narain, K. (1990). New heterotrichid ciliates from *Rana (Dicroglossus) cyanophlyctis. Proc.* 11th *Intl. Congr. Parasit.,* held at Paris, France, from 20-24th August, 1990 (Abt).

Raina, M. K. & Narain, K. (1992). A new Myxosporidian from Gall Bladder of *Puntius chola* in Jammu, with review of the Indian species of *Myxidium* Butschli 1882, from fishes. pp.108-114. *In*: Yousuf, A.R., Raina, M.K. & Qadri, M.Y. (Eds.), *Curr. Trends in Fish and Fisheries Biology, and Aquatic Ecology,* P. G. Department of Zoology. The Univ. of Kashmir, Srinagar, Kashmir (J&K).

Raina, M. K., & Raina, Phoola (1985). On two capillarid nematodes from shrew in Kashmir, with a note on the synonymy of *Skrjabino capillaria* Skarbovitsch, 1946. *Indian J. Helminthol.,* **37**(1):109-115.

Raina, M.K., Sey, O. & Khan, M.D. (1987). Paramphistomes (Trematoda Amphistomida) of domestic ruminants in Kashmir, India. *Misc. Zool. Hung.,* **4:** 5-12.

Raina, R. (1966). *Longidorus reneyii* sp. nov. (Nematoda: Lengidoridae) from Srinagar, Kashmir. *Indian J. Ent.,* **28**(4):438-441.

Ramaseshiah, G. & Dharmadhikari, P. R. (1969). Aphelinid parasites of aphids of Kashmir. *Tech. Bull. Commonw. Inst. Biol. Control.,* **11**:156-164.

Ranga-Rao, G.S.C., Sharma, R.L. & Hemaprasantha (1994). Parasitic infection of Indian Yak, *Bos (Poephagus) grunniens-an* overview. Vet. *Parasitol.,* **53**:75-82.

Rao, K.N.A. (1951). Tick typhus in Srinagar. *Indian J. Med. Res.,* **39**:293-96.

Rao, K.N. A. & Kalra, S.L. (1949). Tick-borne relapsing fever in Kashmir. *Indian J. Med. Res.,* **37**:385-97.

Rao, T.R., Dhanda, V., Bhat, H.R. & Kulkarni, S.M. (1973). A survey of haematophagons arthropods in Western Himalayas, Sikkim and hill districts of West Bengal. A general account. *Indian J. Medical Research,* **61**(10):1421-1461.

Rao, V. P. (1966). Survey for natural enemies of Gypsymoth Report for period 1961-66 – U.S. P.L.480 Project. *Commw. Inst. Biol. Control. Banglore,* India, 50 p.

Rao, V. P., Ghani, M. A., Sankaran, T. & Mathur, K. C. (1971). *A review of the biological control of insects and other pests in South-east Asia and Pacific region. Tec. Communication* No. 6. CIBC, Trinibad. West Indies. Commonwealth Agricultural Bureux. England.

*Rashid, Majidah (1992). *Studies on population dynamics of helminth parasitization of fishes in winter lake under varying habitats and seasons.* Ph. D. Thesis. P. G. Deptt. of Zool., Faculty of Sci., Univ. of Kmr., Sgr., Kmr. (J&K).

Rishi, N. D. & Shah, K. A. (1985). Survey and bioecological studies on natural enemies complex of India. Gypsy Moth, *Lymantria obfuscata* Walker) (Lepidoptera: Lymanpiidae). *Entomol. Res.,* **9**(1):82-93.

Rishi, N. D. & Lone, M. A. (1984). Description of a new species of *Protaphelinus* Mackauer (Hymenoptera: Aphelinidae) parasitic on gall-forming aphid. *Pemphigus bursarius* Linn., infesting *Populus nigra* var. *italica* in Kashmir. *J. Ent. Res.,* **8**(2):132-136.

Rodnigues, F.M., Padbidri, Ghalashi, G.R., Gupta, N.P. Mandke, V.B., Pinto, R.S., Hoon, R.S., Bapat, M.B. & Mohan Rao, C.V. R. (1986). Prevalence of Crimean haemorrhagic fever Congo virus in Jammu & Kashmir State. *Indian J. Medical Res.,* **84**:134-38.

Rognes, K. (1972). A new genus of Halicoboscinae from the Himalaya (Diptera: Calliphoridae), with amended genus and sub-family concept. *Entomologica Scandinavica,* **23**(4): 391-404.

*Sahaf, K. A. (1991). *Studies on the parasitic infection of Bombay mori and their control in J&K.* Ph. D. Thesis, Faculty of Sci., P. G. Deptt. of Zool., The Univ. Kmr., Sgr., Kmr. (J&K).

Sahaf, K. A. (2002). Incidence of Protozoan disease, Pebrine of Silkworm, *Bombyx mori* in Jammu and Kashmir. *J. Entomol. Res.,* **26**(4): 305-307.

Sen, P. (1938). A check and host-list of Ixodoidea (Ticks) occurring in India. *Indian J. Vet. Sci.,* **8**:133.

Sewell, R. B. S. & Roonwal, M. L. (Eds.) (1965). *The Fauna of India and adjacent countries. Diptera., Muscidae,* Part I by Dr. F. I. Van Emiden Published by *Zool. Surv. of India,* Calcutta, W. Bengal, India.

Shah, F.A. (1996). *The Honey Bee and Bee Keeper.* Published by Bee Master, 259 p.

Shah, F. A. & Shah, T. A. 1986 (1988). Search for *Nosema apis* in Kashmir *Apis cerana. Indian Bee J.,* **48**(1-4): 55-56.

Shah, F. A. & Shah, T. A. (1988). Varroa disease of honeybees – A Review. *Indian Bee J.,* **50**: 102-105.

Shah, F. A. & Shah. T. A. (1989). Tropilaelaps disease of honeybees. *Indian Bee J.,* **51**:120-125.

Shah, F.A. & Shah, T.A. 1989 (1991). *Tropilaelaps (Tropilaelaps) clareae)* not a threat to *Apis mellifera* in temperate climate. *Indian Bee. J.,* **15**(3):120-121.

*Shah, Feroz Ahmad (2004). *Histopathological and serological studies of helminth infection in fishes of some lakes of Kashmir.* Ph. D. Thesis. P. G. Deptt. of Zool., Faculty of Sci., Univ. of Kmr., Sgr., Kmr. (J&K).

Shahardar, R. S., Mir, A. S., Mir. M. M., Pandit, B. A. & Ahmad, M. A. (1995). Efficacy of fenbendazole against gastro-intestinal nematodes of Kashmir deer. *Indian Vet. J.,* **72** (7): 781-789.

Shahardar, R. A., Mir, A. S., Pandit, B. A. & Banday, M. A. A. (1993). Successful treatment of goat warbles with inermectin. *Indian Vet. J.,* **70**(7): 666-667.

Shahardar, R. A. & Rashid, A. (1998). Occurrence of larvae of *Oestrus ovis* in goat in Kashmir. *J. Vet. Parasitol.,* **12**(2): 147.

Sharif, M. (1928). A revision of the Indian Ixodidae with reference to the collection in the Indian Museum. *Recd. Ind. Mus.,* **30**:217-344.

Sharif, M. (1930). A revision of Indian Siphonaptera Part I. Family Pulicidae. *Rec. Indian Mus.,* **32**:29-62.

Sharma, B. D. (1972). *Ornithonyssus bursa* (Mesostigmata: Dermanyssidae) on the poultry birds of Kashmir. *Poultry Guide,* **9**(9):52.

Sharma, B. D. (1973). *Psammotettix striatus* (Linn.), a rare hemipteran ectoparasite of the flying foxes of Poonch (J&K). *Curr. Sci.,* **43**(3): 108.

Sharma, B. D. & Das, S. M. (1971). A new record of *Boophilus decoloratus* (Koch) (Acarina: Ixodidae) from Kashmir, India. *Indian J. Entomol.,* **32**(3): 292-293.

Sharma, B. D. & Das, S. M. (1971a). On the occurrence of *Melophagus ovinus* Linn., 1958 (Insecta: Hippoboscidae) on sheep and cattle in Kashmir. *Indian Vet. J.,* **48** (9): 972-973.

Sharma, B. D. & Das, S.M. (1972). A checklist of ticks and their hosts in Kashmir. *Orissa Vet. J.,***7**113.

Sharma, B.D. & Sharma, T. (1972): A new record of two ectoparasitic mites of housefly from J&K State. *Indian J. Anim. Res.,* **6**(1):42.

Sharma, B. D. & Sharma, T. 1974 (1976). A note on contribution of tick-fauna of Jammu Province. *Indian J. Anim. Sci.,* **44**(9):709-712.

Sharma, N. R., Thapa, C. D. & Nath, A. (1981). Pathogenicity and identity of myceliphagous nematodes infesting *Agaricus bisporus* (Lan.) Singh. *Indian J. Nematol.,* **11**:230-231.

Sharma, O. P. 2003 (2004). *Illustrated Dictionary of Biodiversity.* Published by Forest Res. Instt., Jammu & Kashmir, Jammu (J&K), 159 p.

Sharma, R. L. (1994). Parasitic bronchitis in goats and the possible use of *Dactyocaulus filaria* vaccine for its control. *Vet. Parasitol.,* **51**(3/4):255-262.

Sharma, R. L. & Dhar, D. N. (1978). Studies on ovine Oesophagostomoiasis I. Population kinetics following varying levels in lambs. *Proc. Asian Congr. Parasitol.,* p. 300 (Abt).

Sharma, R.L. & Dhar, D.N. (1979). Population kinetics of primary *Oesophagostomum columbianum* infection in lambs. *Indian J. Anim. Sci.,* **49**:203-208.

Sharma, R. L. & Dhar, D. N. (1987). Prevalence of *Dictyocaulus viviparous* infection in Kashmir. *Indian J. Parasitol.,* **11**(1): 101.

Sharma, R.L., Peer, F.U. & Bhat, T.K. (1996). Antiparasitic activity of invermectin against *Psoroptis communis* var. *ovis* infestation in Kashmir Merino sheep. *Indian Vet. J.,* **67**(3):264-66.

Sharma, R.L., Raina, O.K. & Hemaprasanth, 1989 (1990). Post hibernation fate of snail *(Lymnaea auricularia) sensu stricto* experimentally infected with *Fasiola gigantica. Riv. Parassitol.*, **50**(1-1):185-87.

Sharma, V. & Gupta, A. (1985). A new doryctine genus *Protatus* infesting *Sinoxylon* sp., a serious pest of *Dalbergia sissoo* in Jammu (India) (Hymenoptera: Braconidae). *Reichenbachia,* **23**(1):37-39.

*Shaw, A.A. (1971). *Studies on some infections disease and infertility among livestock of Kashmir, with special reference to protozoa*. Ph.D. Thesis. P.G. Deptt of Zool., Univ. of Kmr., Srg.(Kmr.), 301 p.

Shaw, A.A. (1980). Protozoan infection in J&K State. pp. 200-205. In. *Heritage of Kashmir,* Hassnain F.M. (Ed.), Chap.18. Gulshan Publishers, Srinagar, Kashmir.

Shaw, A.A. & Das, S.M. (1970). Bavine Theileriosis in cross-bred Jersey Cow in Kashmir. *Kashmir Sci.,* **7**(1-2):99-102.

Sheikh, A. G. 1977 (1979). Occurrence of *Anguina tritici* (Steinbuck, 1799), Chitwood, 1935, on wheat in Kashmir. *Indian J. Plant Protection,* **5**(2):20.

Sherman, R. A. & Pechter, E.A. (1988). Maggot therapy: a review of therapeutic application to fly larvae in human medicine, especially for treatment of Oesteomyelitis. *Med. & Vet. Entomol.,* **2**: 225-230.

Shuja-uddin (1970). Two new species of Aphidiidae (Hymenoptera) from India. *Rec. Zool. Surv. India,* **68**:415-420.

Shuja-uddin (1974). Three species of *Trioxys* (Haliday) (Hymenoptera: Aphidiidae) recorded from India. *Indian J. Ent.*, **35**:9-14

Shuja-uddin (1976). Two new species of the genus. *Toxares* Westwood (Aphidiidae: Hymenoptera) from India. *Indian J. Entomol.*, **36**:261-274.

Shuja-uddin (1978). New records of *Adialytus* Forster (Hymenoptera: Aphidiidae) from India, with comments on the validity of the genus. *J. Ent.* Res., **2**(2):160-62.

Shuja-uddin (1978a). New Records of *Ephedrus campestris* Stary (Aphidiidae: Hymenoptera). Indian J. Zool., 6(1):60-61.

Shuja-uddin (1983). A new species of the genus *Trioxys,* Haliday (Hymenoptera: Aphidiidae) from Kashmir, India. *J. Ent. Res.*, **7**(1):36-38.

Siddiqi, A. U. & Khan, E. (1983). Taxonomic studies of Tylenchidae (Nematoda) of India. 3. Two new species of *Cephalenchus,* along with description of *Filenchus conicephalus* sp. n. (Nematoda: Tylenchidae) from India. *Indian J. Nematol.,* **13**(1):84-90.

Siddiqi, M. R. (1963). Four new species in the subfamily Tylenchinae (Nematoda) from North India. *Z. fur. Parasitkde.,* **23**:239-244.

Simmonds, F. J. & Rao, V. P. (1960). Record of *Plutella maculipennis* Curt. And some of its parasites in Kashmir, India. *Canad. Ent.,* **92**(4):274.

Singh, H. N., Udaya Lakshmi & Singh, H. N. (1984). Natural parasitization in eggs of *Lymantria obfuscata* at Shrinagar. *Indian J. Entomol.,* **46**(3):365-366.

Singh, H. N. & Udaya Lakshmi, A. (1987). Natural parasitism in eggs of *Lymantria obfuscata* Walker, at Shrinagar, J&K. *Bull. Entomol.,* **28**(1):70-72.

*Singh, Ranvijay (1997). *Ecological and Taxonomic studies on metazoan parasites of some lacustrine fishes in Jammu*, Ph.D.Thesis, Department of Zoology, The University of Jammu, Jammu (J&K)

Singh, Ranvijay, Singh., Gurdarshan & Duda, P. L. 1999 (2001). Community structure of enteric trematode of two sympatric frogs in Jammu. *Oriental Sci.,* **4**(2):57-70.

Sinton. J. (1932). Notes on some Indian species of the genus *Phelbotomus.* xxx. Diagnostic table for the females of the species recorded from India. *Indian. J. Med. Res.,* **20**:55-74.

Sinton. J. (1933). Notes on some Indian species of genus *Phlebotomus*. xxxv. Additions and alterations to the diagnostic table of females. *Indian J. Med. Res.,* **21**: 225-228.

Sinton. J. (1933a). Notes on some Indian species of genus *Phlebotomus.* xxxvi. Diagnostic table for the males of the species recorded from India. *Indian J. Med. Res.,* **21**:417-428.

Smith, F. G. A. M. (1977). Some Siphonaptera of small mammals in Kashmir (Insecta). *Senckenbergiana biol.,* **58**(1-2): 45-53.

Soota, J. D. (1959). Fauna of the Kashmir Valley, Leeches. *Recd. Ind. Mus.,* **54**(1-2):1-4.

Srivastava, S. P. & Wattal, B.L. (1981). Parasitic arthropods of some vertebrates in Gulmarg and Tangmarg areas of Jammu & Kashmir State (India). *J. entomol. Res.,* **5**(2):121-126.

Stary, P. & Bhagat, R.C. (1978). Records and descriptions of aphid parasitoids from Kashmir, India. (Hymenoptera: Aphidiidae). *Act. Ent. Bohemosl.,* **75**:387-393.

Stary, P. & Ghosh, A. K. (1983). *Aphid parasitoids of Indian and adjacent countries* (Hymenoptera: Aphidiidae). *Tec. Monograph* No.7, Edited and published by Director, Zoological Survey of India, Kolkata, West-Bengal, India.

Subba-Rao, B. R. & Hayat, M. (Eds.) (1985). The Chalcidoidea (Insecta: Hymenoptera) of India and adjacent countries. Part-I. Review of families and keys to families and genera. *Oriental Insects,* **19**:163-310.

Subba-Rao, B. R. & Hayat, M. (Eds.) (1986). The Chalcidoidea (Insecta: Hymenoptera) of India and adjacent countries, Part-II. *Oriental Insects,* **20**:1-43

*Sudan,O.S.(1979) *Studies on Tremadode-fauna of Jammu Province.* Ph.D. Thesis, Department of Zoology, Jammu University, Jammu (J& J).

Sudhan, N. A. & Shahardar, R. A. (1999). Occurrence of *Coenurus* cyst in sub-cutaneous tissue of ear of a goat. *J. Vet. Parasitol.,* **13**(1):75-76.

Swarup, G. & Das-Gupta, D. R. (1986). *Plant parasitic nematodes in India— Problem and Progress.* Indian Agriculture Research Institute, New Delhi, India. 497 p.

Swarup, G., Prasad, S. K. & Raski, D. J. (1964). Some *Heterodera* species from India. *Plant Diseases Reporter,* **48**(3):235.

*Tak, Hidayat-ullah (2003). *Studies on epidemiology, pathogenicity and control of hydatid disease in Kashmir Valley.* Ph. D. Thesis, Faculty of Science, P. G. Deptt. of Zool., The Univ. Kmr., Sgr., Kmr. (J&K).

Tak, Hidayat-ullah, Chishti, M. Z. & Ahangar, A.G. (2003). Prevalence of hydatid disease in slaughtered animals in Kashmir Valley, pp. 225-230. *In*: Gupta, Neelan & Gupta, D.K. (Eds.) *Proc. Natl. Congr. Parasitol.,* held at Barielly, U.P. in Feb., 2003.

Takada, Hajimu & Rishi, N.D. (1980). Records of fifteen species of Aphidiidae (Hymenoptera) from Kashmir, India, with descriptions of three new species. *Kontyu,* **48**(2):234-240.

Talwar, P.K. & Jhingran, A.G. (1991). *Inland Fishes of India and Adjacent. Countries.* Vol. I-II. Oxford and IBH Publishing House, New Delhi. Vol.I, 1-541 pp., Vol.II, 542-1158 pp.

*Tanveer, S. (1996). *Biology and pathogenicity of parasites of some poultry and economically important birds of Kashmir.* Ph. D. Thesis. P. G. Deptt. of Zool., Faculty of Sci., Univ. of Kmr., Sgr., Kmr. (J&K).

Tanveer, S. and Chishti, M. Z. (1999). Studies on some strongylid and trichostrongylid nematodes in avian hosts of Kashmir. *Oriental Sci.,* **4**(2):93-100.

Tanveer, Syed & Chishti, M.Z. (2001). Studies on Notocotylid trematode genus *Paramonostomum* (Leuhe) 1909 in domestic fowl and common coot in Kashmir, with description of a new species. *J. Parasitic Diseases,* **25**(2):95-99.

Thapar, G.S. (1950). Two new species of the genus *Rhabdochona* Railliet from Indian Fishes *Indian J. Helminthol.,* **2**(1):35-40.

Theodor, O. (1948). Classification of the old world species of the sub-family Phlebotominae (Diptera: Psychodidae) (illustrated). *Bull. Entomol. Res.,* 85-115.

Thokar, Manzoor, A. (1986). *Comparative evaluation of enzyme linked immunosorbent Assay for detection of Toxoplasma gondii antibody in clinically suspected toxoplasmosis.* M. D. Thesis Department of Microbiology, Shir-e-Kashmir Institute of Medical Sciences, Srinagar, Kashmir. p. 69.

*Tickoo, Ratni (1969). *Studies on the morphology and life-cycle of some helminth parasites of amphibians in Kashmir.* Ph. D. Thesis. P. G. Deptt. of Zool., Faculty of Sci., Univ. of Kmr., Sgr., Kmr. (J&K).

Tickoo, Ratni (1970). On the morphology of *Haematoloechus almorai* (Pande, 1937) from the lungs of *Rana cyanophlyctis* in Kashmir. Part I: General Morphology cuticle and musculature. *Kashmir Sci.,* **7**(1-2):156-157.

Tripathi, Y.R. (1956). Studies on parasites of Indian fishes. *V. Acanthocephala. Rec. Indian Mus.,* **54**:61-99.

Trivedi, P. C. 1998 (Ed.). *Phytonematology in India.* CBS Publishers and Distributors, Daryaganj. New Delhi, India, 205 p.

Van Cleave, H.J. (1948). Expanding horizon in the recognition of Phylum. *J. Parasitol.,* **34**:1-20.

*Wali, Nirja (1983). *Studies on pathogenic nematodes of sheep.* Ph. D. Thesis P. G. Deptt. of Zool., Faculty of Sci., Univ. of Kmr., Sgr., Kmr. (J&K).

Wali-ullah, M. I. S. (1992). Nematodes associated with vegetable crops in Kashmir Valley, India. *Nematol. Medit.,* **20**:47-48.

Wali-ullah, M. I. S. & Bhat, M. A. (1990). Occurrence of mixed population of *Pratylenchus* Filepjer, 1936 on potato *(Solanum tuberosum)* in Srinagar, Kashmir. *Indian J. Nematol.,* **20**(2):239-241

Wali-ullah, M. I. S., Masoodi, M. A. & Kaul, V. (1993). Population behaviour of nematode associated with strawberry *(Fragaria vesca)* in Srinagar, Kashmir. *Curr. Nematol.,* **4**:41-44.

Wali-ullah, M. I. S. & Kaul, V. (1997). Nematodes associated with cherry plant in Kashmir Valley. *Indian J. Nematol.,* **27**(2):237.

Wali-ullah, M. I. S., Kaul, V. & Zaki, F. A. (1991): The Rice-root nematode, *Hirschmanniella* spp. In Jammu and Kashmir. *Geobios new Report,* **10**:175-176.

Wani, B. A. (1970). *Incidence and Symptomatology of ascariasis in Kashmir.* M. D. Thesis (Medicine). Govt. Medical College, Srinagar. The Univ. of Kashmir, Srinagar, Kashmir (J&K).

Wani, N. A., Babu, M. L., Ahangar, A. G., Kirmani, A. R. & Saxena, R. K. (1999). Hydatid cyst of Brain —A case Report and Review of literature. *JK–Practitioner,* **6**(2): 145-146.

Wani, N. A. & Chrungoo, R. K. (1992). Biliary Ascariasis: Surgical aspect. *World J. of Surgery,* **16**:976-79.

Wattal, Chand, Malla, Nancy, Khan, Imtiyaz & Agarwal, S.C. (1986). *In vitro* development and hatching of *Ascaris lumbricoides var. hominis* egg. *Indian J. Med. Res.,* **83**:257-264.

White, S. R., Aubertin, D. & Smart, J. (1940). *The Fauna of British India, including the remainder of the Oriental region,* Vol.VI. Calliphoridae. Taylor and Francies, London, England.

Yamaguti, S. (1958). *Systema Helminthum* Vol.I (Part I & II). *The Digenetic Trematodes of Vertebrates,* Vol. IV. *Monogenea and Aspidocotylea* Interscience Publishers Inc. Indian Edition (Reprint) 1985, International Books and Periodical Supply Co., New Delhi, India.

Yamaguti, S. (1959). *Systema Helminthum* Vol.II. The *Cestodes of Vertebrates*. John Wiley & Sons, Japan.

Yamaguti, Satyu, 1963. *Systema Helminthum,* Vol. V. *Acanthocephala*. John Wiley & Sons. Inc. Indian Reprint Edition, 1985, International Books and Periodical Supply Service, New Delhi, 423 p.

Yamaguti, S. (1971). *Synopsis of Digenetic trematodes of Vertebraters,* Vol. I. Keigaku. Publishing Co. Tokyo-Japan.

Yamaguti, Satyu (1985). *Systema Helminthum* Vol. III, Part I & II, *Nematodes of Vertebrates,* Indian Reprint Edition, Part I, 1-679 pp. and Part II 681-126 pp. International Books and Periodical Supply Service, New Delhi, India.

Yaseen, M. (1979). *Intestinal functions of Ascariasis.* M.D. Thesis (Medicine). Govt. Medical College, Srinagar. The Univ. of Kashmir, Srinagar, Kashmir (J&K), 55 pp. (unpublished).

Yattoo, G.N., Khuroo, M.S. Zargar, S.A., Bhat, F.A. & Sofi, B.A. (1999). Case Report: Precutaneous drainage of the pancreatic head hydatid cyst with obstructive jaundice. *J. Gastroenterology & Hepatology,* **14**(9):931-34.

Zaka-ur-Rab 1972(1974). Record of *Syrphus confrater* Wiedmann as predaceous on woolly aphid, *Eriosoma lanigerum (*Husum*). Indian J. Ent.,* **34**(4):348-49.

Zaki, F. A. (1999). A note in Ladakh (J&K). *Appl. Biol. Res.,* **1**(1):175-177.

Zaki, F. A., Kaul, V. & Wali-ullah, M. I. S. (1991). Plant-parasitic nematodes associated with potato-turnip rotation and other vegetable crops in Kashmir. *Indian J. Nematol.,* **21**(2):174-75.

Zaki, F. A., Kaul, V. & Wali-ullah (1993). Crop losses estimation in tomato and potato caused by plant-parasitic nematodes in Kashmir. *Indian J. Nematol.,* **23**(1):101-104.

Zaki, F. A. & Mir, A. A. (1994). First record of ear cockle nematode *(Anguina tritici)* on wheat in Ladkah. *Indian J. Nematol.,* **24**:228-29.

Zeya, S. B. & Hayat, M. (1995). A revision of the Indian species of *Gonatocerus* Nees. (Hymenoptera: Chalcidoidea: Mymaridea). *Oriental Inst.,* **29**:47-160.

Zutshi, M. K. (1967). *Lymantria obfuscata* Walker and its natural enemies in Kashmir (Lymantridae: Lepidoplera). *J. Bombay Nat. Hist. Soc.,* **64**(1):126-129.

***Denotes author and title of the awarded Ph.D. theses, with study area as Jammu & Kashmir and not cited in the text.**

AUTHOR INDEX

E

F

G

H

I

J

K

L

M

N

O

P

U

V

W

Y

Z

SUBJECT INDEX

F

G

H

I

K

L

O

P

W